SOME CALL THEM SUPERMEN, OTHERS CALL THEM GODS

Actually, we do not yet know the exact nature of the Intelligences from outer space who have for millions of years visited Earth and shaped her history.

But as more and more detailed evidence comes to light, we are becoming sure that these Intelligences have played and continue to play a vastly important role in the destiny of Earth and humankind.

With this brilliant, bold, yet completely scientific book, you, the reader, are at last offered the proven facts that can fuel your mind for one of the most daring and important flights of vision that you have ever experienced.

Extraterrestrial VISITATIONS *from Prehistoric Times to the Present*

by

Jacques Bergier

A SIGNET BOOK
NEW AMERICAN LIBRARY
TIMES MIRROR

First published in France under the title *Les Extra-Terrestres dans l'Histoire* by Editions J'ai Lu.

Library of Congress Catalog Card Number: 73-6451

This is an authorized reprint of a hardcover edition published by the Henry Regnery Company.

Signet, Signet Classics, Mentor, Plume and Meridian Books *are published by The New American Library, Inc., 1301 Avenue of the Americas, New York, New York 10019*

First Printing, July, 1974

1 2 3 4 5 6 7 8 9

PRINTED IN THE UNITED STATES OF AMERICA

Contents

Prologue

Plea for an Open History

The concept of a closed history is relatively recent. By closed history I mean a history in which all events are considered the result of natural or human causes. Throughout most of its past, humanity has also believed in the intervention of external causes in history: demons, supernatural creatures, gods, and, finally, God. It was not until the nineteenth century that the idea of a history totally without external intervention, one in which causality was exclusively limited to our own planet, gained credence. And, like so many ideas of the nineteenth century, this is susceptible to argument and has in fact led to a great deal of debate.

The purpose of this book is to spot those external interventions, in prehistory as well as in history, the origin of which cannot possibly be restricted to our planet.

This book maintains an exclusively rationalist position. The interventions I discuss have been the acts of intelligent beings more advanced than ourselves: physical beings living in space.

I will not be talking about so-called supernatural interventions (each person has the right to his own opinion on that subject). Nor will I be talking about "flying saucers," which have already been much discussed. Nor do I pretend to supply absolute proof of interventions by extraterrestrial beings in the course of our planet's prehistory and history. Other researchers, with tools available to them that are superior to mine, will surely do this before the end of the century.

I would rather compare myself to those eccentric characters who, before the appearance of *Origin of Species*, published "bizarre" books—from which Darwin learned. I would be quite satisfied if my book succeeded in interesting a large number of readers and if among those readers there was some Darwin of the future to whom I had given the desire to research further.

For me, having extraterrestrial beings intervene in our history is no more absurd than having microbes intervene in the state of our health. In both cases, it is a matter of interventions that are imperceptible to our senses but are revealed by deeper study and confirmed by instrumental analysis. Thus, I think, the study of the strange facts that I have gathered here will one day permit us to establish the fact of intervention of beings who have come from the outside to modify the course of our history.

Charles Fort said: "We are the property of someone." I go further than he does, in declaring that we are the creation of someone; and less far, in postulating that we are under surveillance and that perhaps "they" intervene in our activities and in our history.

Why hasn't there been, and isn't there, direct, open contact between "them" and us? This question is frequently debated. For my part, I believe that these contacts do exist, but that they are hidden from man as a whole and have taken place, at well-defined periods, only with very advanced individuals who have been above the average of their fellows.

Legends concerning these contacts are certainly at the base of numerous traditional stories. But since there is no formal proof in this area, I have preferred to limit myself here to studying contacts in the extraterrestrial-earth direction, that is, one-way contact originating outside earth. And, as we will see, even so limited an approach offers material of great interest and further elaboration.

1

The Star that Killed the Dinosaurs

Seventy million years ago, the earth was inhabited by giant reptiles: gigantic lizards, colossal saurians, who slithered, swam, flew. Their reign lasted one hundred million years—whereas, according to the most optimistic estimates, man has had barely six million years.

This means that these species of reptiles had in order to become adapted and to evolve, an infinitely longer time than man. Furthermore, it is impossible to pretend that they represented an evolutionary failure: any species that lasts a hundred million years must be considered fully adapted. Yet few species that were contemporaries of those reptiles survive—for example, certain crabs, which have not changed in three hundred million years. In fact, in less than one million years the giant reptiles entirely disappeared.

How and why?

We can scarcely maintain that it was because of a change in climate; for even when the climate changes, the oceans hardly vary, and many of these reptiles lived in the oceans.

It is impossible to believe that a higher form of life was able to exterminate them. This would have required a considerable army, whose traces we would certainly have found.

One amusing hypothesis is that our ancestors, the mammals, might have fed on dinosaur eggs. But it is only that: an amusing hypothesis: the icthyosaurs deposited their eggs in the oceans, out of their adversaries' reach.

It has been said that the grasses changed, and that the new grasses were too tough for the big reptiles. A completely unlikely hypothesis: large numbers of vegetation types survived, on which they could have fed perfectly well. The giant tortoises of the Galápagos Islands, the ones that interested Darwin so much, did not die of hunger.

One could say that species grow old, become senile, and die. But this is bad logic: the preservation of the genetic code prevents a species from dying out. And why haven't those species that are still living after several hundred millions of years, such as crabs and cockroaches, become senile too?

None of these hypotheses hold. But *something* happened. What then? An ingenious hypothesis has been outlined by two Soviet scientists, V. I. Krasovkii and I. S. Chklovski, both of whom are eminent astrophysicists—especially the latter, who is the author of some extremely important works in astrophysics and radio astronomy. It was Chklovski, in fact, who studied synchotron radiation and showed that relatively rapid and extremely violent events can be produced at the center of galaxies as well as in space in general.

The two Soviet scientists explain the end of the dinosaurs by hypothesizing a star explosion that occurred at a relatively small distance from our solar system: a supernova at five or ten parsecs from us that would have increased the density of radiations coming from space. Lending credibility to this theory, the English radio astronomer Hanbury Brown believes he has detected traces of the explosion of a supernova fifty thousand years ago at a distance of only forty parsecs from the solar system.

Two U.S. scientists, K. D. Terry of the University of Kansas and W. H. Tucker of Rice University, have recently given close quantitative study to the problem. They have observed stars that, when they explode, actually produce such radiation bombardments. The effect of a bombardment varies according to the intensity of the earth's magnetic field. This field partially protects us from the bombardment of cosmic particles by turning away those with a magnetic charge and forcing them into orbit around our planet. But this magnetic field varies in intensity. Right now it is on a downswing and will reach a lowpoint about the thirty-sixth century A.D. It is possible that seventy million years ago a violent bombardment may have coincided with a diminution in the earth's magnetic field, bringing about a wave of mutations in which the dinosaurs died and man's ancestors were born.

According to an East German scientist, Richter,

the bombardment originated at the center of our galaxy and was extremely powerful, in spite of being produced at such a considerable distance.

If we accept this theory, as we well may, we still must ask: What caused the massive explosion? I first outlined my explanation in 1957 during a broadcast on French television. I still remember the uproar that followed, and I still stand by my hypothesis: that the star explosion that killed the dinosaurs was deliberately induced, designed to set off a slow process of evolution leading to intelligent life; that we were created by extremely powerful beings. Knowledgeable both of the laws of physics and of the laws of genetics, these beings —who could truly be called gods—set in motion a series of events that will not stop with man but will continue until this evolution results in other gods, beings equal to their creators.

Obviously, this hypothesis is very bold. More than once already, however, we have speculated on the existence of beings infinitely more powerful than ourselves. We have even offered quantitative estimates on what their technologies might be.

The greatest source of energy, which is demonstrated in the H-bomb, is the conversion of hydrogen to helium. Now the amount of hydrogen in the oceans is enormous, but there is even more in the sun. We can certainly imagine beings capable of extracting hydrogen from the sun and using it. Theoreticians call the civilizations of such beings civilizations of type III.

What has become of these civilizations? Do they still exist in the universe?

Many fine minds are answering this question affirmatively. Chklovski considers it a not entirely far-fetched hypothesis that the quasars and the pulsars, up to now unexplained celestial objects, may be signs of biological activity. The Soviet scientist believes we should examine the sky systematically for what he calls "miracles"; that is, phenomena that cannot be explained solely by known natural laws or by imaginative extrapolations of those laws.

Among these phenomena, Chklovski would place:

—the abnormal behavior of Phobos, the satellite of Mars that, according to him, might be an artificial construction;

—the observation of a particular type of star, type *R*, that produces a short-lived element that does not exist in nature—technetium. This condition has suggested to eminent scientists the possibility that intelligent beings are bombarding these type *R* stars with technetium to produce a signal.

Other serious investigators—Carl Sagan, for example—believe that beings of type III civilizations can modulate the electromagnetic emissions of a star as easily as we modulate those of a transmitter of sounds and pictures, and a group of Soviet researchers under the direction of astrophysicists Kardaschev and Pschenko are currently investigating such signals.

This group believes that, considering the various static disturbances that affect the earth's atmosphere, it will be necessary to place a radio observatory on the dark side of the moon in order to detect these signals. But in any case, their calculations have demonstrated to them that with completely conceivable means of energy, signals can be sent up a distance of 13,000 parsecs—in other words, to a greater distance than that which separates us from the center of the galaxy.

A U.S. astrophysicist, Freeman J. Dyson, envisages an even more fantastic "miracle" in the sky. He believes that there exist beings who can utilize the entirety of energy produced by their star. These beings must no longer be living on planets; instead, they inhabit artificial spheres that they have built themselves and that totally surround their star.

Other "miracles," it seems, have been observed in the sky. Though they are rarely mentioned in scientific publications because they seem too fantastic, you hear them talked about at astronomical congresses, during meals and in the corridors. For example, there is speculation concerning certain multiple star systems that might be composed of stars of different ages; that is to say, stars coupled *necessarily* as a result of intelligent activity.

We see, therefore, that the existence in the universe of beings much more powerful than ourselves is being taken as a serious possibility. It is being quantitatively envisaged by eminent scientists.

We need not add to these speculations. We should simply note that the human form perhaps ought not to be rejected. It is possible that it is one of the major stable forms of intelligence in the universe. We do not utilize one-tenth of our brains. Our civilization is far from being perfect. We do not have the least idea of what a civilization would be like in which human beings used 100 percent of their brain capacity. And it is not at all absurd to attribute to such a civilization in the stars, if it exists, powers analogous to those of type III civilizations.

Such a point of view seems more plausible to me than all the inventions of science fiction. But be that as it may, my intention is not to study the possible forms of extraterrestrial beings but rather what I believe to be their *manifestations.*

To me, the first of these manifestations might have had as its result the end of the dinosaurs. Considering that the earth's evolution had been heading down a dead end, that there had not been any progress with the giant reptiles for fifty million years, the Intelligences, wanting perhaps to increase the number of their "brothers in reason" (an expression of the great Soviet mathematician Kolmogoroff), reversed the direction of this evolution and set up a new evolutionary goal. We do not, for the moment, know to what it is leading; but it would be absurd to believe that man as he is today was the end they were pursuing.

Perhaps the Intelligences will be forced to wipe out our species and set another chain in motion. This may be an inspired intuition. In any case, the Intelligences seem far removed from H. P. Lovecraft's Great Old Men, who created life on the earth by mistake or as a joke.

With this concept in mind, it is interesting to note that intelligent signals have been spotted coming from celestial object CTA 102. At least the wavelength of these signals has been calculated as coming from the vicinity of CTA 102, and it has been established that they contain the fundamental wavelength of the universe, the radiation transmitted by interstellar hydrogen.

Since that discovery, Professor Gerald Feinberg of the New York Academy of Sciences has advanced the opinion that these signals from the Intelligences are transmitted by particles that he calls *tachyons*, from a Greek word meaning *rapid*—particles that can be conveyed in the void more rapidly than light without any contradiction of Einstein's theory.*

When methods for observing and detecting the tachyons are perfected, we will probably receive signals being transmitted by other races whose evolution also was set in motion by the star that killed the dinosaurs. And perhaps we will detect the observation devices of the Intelligences, who are certainly observing us in the same way we observe microbe cultures under a microscope.

Although we have some precise notions on the energy sources the Intelligences might have had available for making their experiments—sources that are extensions of those we ourselves are using in our experiments on the hydrogen and antimatter bomb—our knowledge in the area of genetics is still too vague for us to be able to

*This comes from the fact that their mass at rest is measured by a purely imaginary number. It is unfortunately impossible to go into the details without using mathematics. However, this discovery is as important as that of the atomic bomb or of the transistor crystal.

imagine how directed mutations could be produced at a distance. We first need to know how mutations can be directed.

In order to produce directed mutations, a radiation of very short wavelength, or particles with very high energy, would have to be employed.* It would next be necessary to modulate this transmission in order to transfer genetic characteristics on this modulation in the same way that images are transferred by a television channel. Calculations will show that this is possible, provided that very short waves, on the gamma ray band, or particles on the Broglie wavelength and on the short gammas, are used.

Indeed, it is possible for us to conceive of a laboratory apparatus—for example, a laser—regulated to produce the kind of radiations that would induce mutations. But when we try to imagine a mutation on the scale of a star created and modulated by the Intelligences, our imaginations fail. A virus would undoubtedly have the same difficulty trying to imagine a laser.

The creatures who built and modulated the star that killed the dinosaurs really were gods. They are probably the same Intelligences who regulate to a hundred billionth the frequency of those celestial objects called pulsars, and the fact is that more and more scientists admit, in private, that the pulsars are artificial—so many, indeed, that

*These are the same things, the wavelength accompanying a particle being inversely proportional to its energy, according to Dr. Broglie's calculations.

this hypothesis will probably soon appear publicly in most official publications.

Having noted this, we must recognize that most signs of the Intelligences' activity must escape us for the moment. In the same way that whole civilizations have lived without knowing about the radio, or of the existence of other solar systems, we will very probably remain unaware of highly important phenomena that, were we able to detect them, would undoubtedly prove to us the existence of other civilizations.

It is equally possible that perfectly classic phenomena may in reality be, without our suspecting it, manifestations of intelligent activity. In this regard, I can cite two hypotheses.

John W. Campbell, the late physicist and science-fiction writer, studied cosmic particles—particles reaching us from space endowed with very high levels of energy capable of reaching 10^{17} electron-volts. These particles are made of element groups with which we are familiar, ranging from hydrogen to iron. But this very ordinary matter is launched abruptly at a formidable speed, as if a fraction of interstellar gas had suddenly been accelerated until it reached speeds bordering the speed of light. The mechanism that produced this acceleration remains unknown, although Fermi, Chklovski, and many other scientists have offered different models.

Campbell, for his part, suggested that the universe is full of spaceships that are moving at speeds close to that of light. As these ships sweep interstellar gas before them, we observe a trail—

a wake—that is none other than the cosmic rays we have detected.

It cannot be said that Campbell's hypothesis has exactly been received with delirious enthusiasm by physicists. However, one American physicist, Robert Bussard, has suggested a model for an interstellar ship that would absorb interstellar gas by means of a scoop placed in the bow, from which it would obtain energy through fusion, and would then utilize the products of the reactions—such as propulsion fluid—the propellant then being ejected through the stern.

If the universe is full of this sort of spaceship linking the stars, then Campbell was right. And that happened to him almost exasperatingly often.

We can also imagine perfectly well that the mysterious variable objects called pulsars are beacons guiding these interstellar ships in the night of space.

A second hypothesis comes from a Russian scientific writer, Ekaterina Zouravleva. According to Mme. Zouravleva, we constantly receive signals from space, signals that were sent at the birth of humanity and undoubtedly well before that. This signal is made up of the aurora borealis and the aurora australis (the northern and southern lights).

Whatever the truth or otherwise of these two examples, the principle of the hypotheses is probably good even if their authors do not take themselves too seriously. Chklovski had cause to observe one day, in a conversation with friends, that there are two kinds of hypotheses: the working hypothesis, intended to serve as a point of departure for study; and the conversational hypothesis, which serves to pass the time agreeably between two meetings on the mathematics of interstellar plasma.

The basic hypothesis of this chapter—that we are the result of a series of mutations set off from the outside—is a working hypothesis; the presence of messages to be decoded in an aurora borealis is a conversational hypothesis.

What other solar systems, or at least what other stars possessing solar systems, could have been influenced by the artificial source of energy that killed the dinosaurs?

If we go a reasonable distance from our solar system—for instance, fifteen light years—we find five: Alpha Centauri, Epsilon Eridani, 61 Cygni A, Epsilon Indi, and Tau Ceti. In centuries to come, research will undoubtedly be carried out to see if life exists in these systems. If it does, it will be interesting to find out if this life resembles ours, if the rocks on the planets of these systems bear traces of a cosmic bombardment that was produced, on our scale, seventy million years ago. With such a second fix, we could then locate in space the star, either artificial or artificially controlled, that killed the dinosaurs.

Until we have such a cross-reference, unfortunately, we are unlikely to discover our "founding" star. Within a distance of a thousand light years, we find approximately ten million stars, and it is currently impossible to know which of these might be the dead debris of an artificial star created by the Intelligences.

The destruction of the dinosaurs certainly came from the cosmos and not from our solar system, but the study of cosmic influences connected to the galaxy is still in its infancy. We have been able to observe numerical coincidences, which are perhaps only coincidences. (For instance, the frequency of the great glacial periods, about two hundred fifty million years, corresponds roughly to the rotation period of our solar system around the center of the galaxy, which is about two hundred thirty million years.) And an attempt is also being made to determine the frequency at which the center of our galaxy, where chain star explosions and disturbances are produced of which we have only a very faint idea, throws off sprays of condensed matter.

Unfortunately, we still have far to go in this study, though interesting discoveries are being made. For example, it is being debated whether or not it is the chain star explosions that are at the origin of the mysterious quasars, which are scarcely more bulky than single stars and which liberate as much energy as whole galaxies. It is generally admitted that these quasars are something completely new and that it is currently impossible to envisage a scientific hypothesis that could render an account of them.

Some scientists think that humanity might one day be able to understand and even reproduce the energy source of these quasars. This is one of the justifications for the fantastic budgets swallowed up by such organizations as CERN (European Center for Nuclear Research). For myself, I feel that humanity is already menaced with destruction by the H-bomb and that a brake must be put on research institutes that might put at our disposal fantastic powers for whose use the human race is not yet ready. The ancient alchemists were completely right in believing that the secrets of matter had to be jealously guarded. If Hitler had had the means of exploding a star the way the Intelligences exploded the star that killed the dinosaurs, he would certainly have done so. Therefore I hope that studies of very high energies will not succeed for some time and that the power to kindle and extinguish stars freely will never be entrusted to the military or to politicians.

We have already made, in going from TNT to the H-bomb, a jump of 10^7. What this means is that a hydrogen bomb weighing one ton can liberate an amount of energy equal to that of ten million tons of TNT. This is what we call a ten-megaton bomb, and of course bombs of this sort actually exist. A comparable jump would take us from the energy of the H-bomb to the energy necessary for inducing the explosion of a star. In other words, a progress similar to that which brought us in twenty years from the TNT bomb to the H-bomb would place the power of the Intelligences at our disposal.

Thus I hope that this development will not happen during the lifetime of current humanity, which has proved only too well of what it is capable. But to believe that this progress will *never* occur, in a universe that has existed at least twenty billion years, is extremely naive.

Somewhere in the universe, viruses evolved into Intelligences. If the phenomenon occurred several times, the different Intelligences must have come into contact; as Teilhard de Chardin said, "Everything that rises must converge." A botanist at Harvard, Elso Barghoorn, has proved that certain bacteria lived on the earth three billion years ago. It has required this time, and the assistance of the Intelligences, to lead from these bacteria to us, and even if it takes ten billion years for the Intelligences to appear naturally, this time would only be half the observable age of the universe.

There is nothing in known science that conflicts with theories concerning the existence of the Intelligences. There is no longer anything conflicting with the possibility that they may have intervened.

Perhaps they have set up a detection and observation satellite in our solar system that is none other than the mysterious Phobos, the satellite of the planet Mars. Perhaps they have set up those protective radiation belts around the earth that we are just beginning to discover.

Perhaps . . .

In the following chapters we will quickly pull together known history, and then the present, looking for other traces of outside intervention.

2

Dr. Gurlt's Cube

A few years ago, the famous Soviet scientific journalist G. N. Ostroumov went to the Salzburg Museum and asked permission to examine a cube, or rather a parallelepiped, discovered by Dr. Gurlt in a coal mine in the nineteenth century. According to several nineteenth-century investigators, this object, discovered inside a coal bed that was several million years old, had nonetheless been machine-made.

The journalist did not find the cube and felt that the museum officials had treated him rather badly. They told him that the object probably had been lost some time before World War II and that there was not even any normal proof of its existence.

Ostroumov, furious, later published a series of articles in which he asserted that the object was a hoax.

Given that we have indisputable descriptions that appeared in the nineteenth century, the Soviet journalist's assertions about Dr. Gurlt's cube are obviously exaggerated. Still, it would be very interesting to examine the doctor's find with modern scientific methods: for of course millions of years ago there was no industrial civilization on the *earth.*

Equally, the fact is that over the years a certain number of objects of this type, some cylindrical, others having angles of intersection, have been discovered, and that although there may be an explanation for the cylindrical objects, the objects with intersecting angles seem to have been left on the earth by *visitors.*

But before we get to that point, we should consider two little-known incidents, the authenticity of which admits of no doubt.

The first incident: in the fall of 1868, in a coal mine near Hammondsville, Ohio, a miner named James Parsons was working relatively close to the surface. Suddenly, a large amount of coal fell into the shaft, unmasking a slate wall covered with inscriptions. A crowd of people soon gathered and schoolteachers from the area noted a resemblance between these inscriptions and Egyptian hieroglyphs. Given the age of the coal vein, the inscriptions had to date back at least two million years. The inscriptions oxidized too quickly for the big-city experts who came to examine them to decipher them in time.

The second incident: on February 2, 1958, in a uranium mine in the state of Utah, four miners, Charles North, Ted MacFarland, Tom North, and Charles North, Jr., were dynamiting a fossilized tree located in the middle of a bed of high-grade uranium ore. The explosion shattered the tree trunk, revealing a cavity. Inside this cavity—a living toad! The toad lived for another twenty-eight hours. He was thin, but for a creature several million years old, he was in rather good shape.

Incidents of this sort number into the thousands, all perfectly authentic; indeed, the working of mines sometimes holds the possibility of discoveries as important as, or even more important than, those made by archeological exploration. Among these discoveries have been, in the United States, England, Germany, Italy, and France, a large number of metallic objects, some cylindrical and others with intersecting angles, that appear to be made of iron. In the case of the cylindrical objects, the question was apparently solved last year in the USSR, though the solution was arrived at under unusual circumstances that require a preliminary explanation.

In the Soviet Union, the thesis of the present work comes under the category of "counter-faith" theses. In the USSR, all sorts of unexplained phenomena are explained as the acts of *Pricheltzy*—the Russian name for cosmic visitors. The government maintains that such explanations allow it to combat religion. This is debatable; it is not really clear that this allocation of cause does not actually reinforce religion further. In any case, this position has been responsible for sensitizing the minds of large numbers of workers to the interventions of extraterrestrial beings.

This is why, when a cylindrical iron object was

found buried in a Ural Mountains coal bed several millions of years old, the Academy of Sciences was immediately informed. The miners who made the discovery examined the object carefully without damaging it and then advised the local section of the League Against Religion, who then notified the Academy. The object was transported to the University of Moscow as meticulously as if it had come from the moon.

And there Russian scientists proceeded to study it. The object was indeed iron and cylindrical. But detailed studies of cuttings made with a diamond saw showed that it was the branch of a petrified tree, in which the action of certain extremely rare microbes had transformed the calcium into iron.

The Ural Mountains discovery thus seems to furnish the explanation for the cylindrical objects. Unfortunately, none of the objects with intersecting angles has been placed under similar examination. These objects most often belong to private collectors, who refuse to entrust them to scientists. And, lacking a study actually demonstrating the contrary, we may still at the present time admit the possibility that these angled objects have come from the outside and were not manufactured on earth. This is the hypothesis I shall maintain in the remainder of this chapter.

What could these objects be? Why were they deposited on our planet when the plants that were later carbonized were still growing? The answer to the first question will give us the answer to the second.

In my opinion, they were *data collectors* of the same type as magnetic bands, but much more highly perfected.

Detailed calculations have been made on the possibilities of a data collector made of iron and having the capacity of a human brain. The results are surprising.

If we assume 100 percent efficiency in data storage and retrieval, duplicating the contents of a human brain would require an iron cube with 2.10^{10} atoms—or to put it another way, a cube of one thousandth of a millimeter, smaller than the head of a pin. One hundred percent efficient cubes or parallelepipeds measuring several hundred centimeters per side could store the most highly detailed data on everything that has taken place on our planet during the last ten million years!

The data could have been fed to these collectors by radiations of which we were unaware and which scanned our planet like radar. If all this is so, then one day these objects will no doubt disappear from the museums just as the Salzburg cube disappeared: they will have been retrieved by the Intelligences who placed them on the earth.

This is far from being science fiction. The hypothesis follows far too logical a line of reasoning. If life on earth was artificially modified, the experiment must have been followed up, and from time to time "they" must have retrieved the recording devices that were placed on the earth during the seventy million years since their experiment was begun.

Clearly it would be extremely useful to trace

down all those angled objects discovered in mines and to note those that have mysteriously disappeared. Even more interesting will be the discovery of one of these objects, in the "magnetic fields" of which we will find stored data on the periods prior to the appearance of man. The existence of such recording devices, on earth or in the earth's environs, on artificial satellites built by beings other than man and older than man, seems fairly certain to me.

For slightly less than a century now, life on earth could have been detected through the radio waves we have been transmitting and which must surely have reached other civilizations. Before our discovery of radio waves, however, events on earth could only have been followed by devices analogous to radar; it is tempting to believe that the results of such exploration have been recorded on the earth itself, and that the recording devices were later retrieved.

Some persons think that the famous iron pillar of Delhi could be such a recorder. The Delhi pillar is eighteen feet high and has a diameter of eighteen inches. It bears the inscription of an epitaph of King Chandragupta II, who died in 413 A.D. So far as is known, the column was already quite old at the time.

Now techniques for manufacturing steel existed in India; one of the princes from the Punjab presented Alexander the Great with a steel ingot weighing some five hundred pounds, an enormous amount for that period. On the other hand, alchemy was highly developed and iron is the essential metal of alchemists. And yet . . .

And yet, considering the extraordinary quality of the metal of which the pillar is made, considering that it is being preserved indefinitely, I still have to ask if it is not a giant recorder. Magnetic analysis could be most revealing, but, unfortunately, in view of the sacred value attached to the iron pillar of Delhi, it is not possible to perform this experiment.

The hypothesis that the pillar is a recorder is plausible, since the various explanations made of this pillar that never oxidizes, even during the rainy seasons, are totally inadequate. Privately published explanations that I have seen suggesting that the pillar was manufactured with the use of blasting metallurgy, amount in my opinion to a display of complete ignorance of that technology. To create an object of this size through fritting would require molds and treatment furnaces of far greater dimensions than any that have been achieved at the present time. It is difficult to believe that installations of this size could have been built in the past. And it is even more difficult to believe that no traces of them have remained.

Perhaps the explanation is made easier if we return to the story of Dr. Gurlt's cube. Dr. Gurlt found the cube in 1865, in a coal mine in Germany where it was deeply embedded in a layer dating from the tertiary. It had been there for tens of millions of years, undoubtedly from shortly after the dinosaurs met their demise. In 1886, Dr. Gurlt made his find public.

Several other works on this subject also appeared, notably in the transactions of the Academy of Sciences. The object was nearly a cube, with two opposing faces of the cube being rounded slightly. It measured about two and a half inches by one and four-fifths inches, the latter measurement being taken between the two rounded surfaces, and weighed about twenty-eight ounces. A fairly deep incision went all the way around about midway up its height. Its composition was that of hard carbon-and-nickel steel. It did not have enough sulfur to be made of pyrite, a natural mineral that sometimes takes geometric shapes.

Some specialists of that period, including Dr. Gurlt himself, said that it was a fossil meteorite. Others declared it a meteorite that had been reworked. But by whom? By the dinosaurs?

Finally, still other experts said that the object was artificially manufactured, and it is this that I think likely.

In any case, the object was deposited in the Salzburg Museum, and there was less and less talk of it. In 1910, it no longer appeared in the museum's inventory. Where did it go? No one knew anything about it.

Between the two world wars, the museum authorities, no doubt annoyed by the number of questions put to them about it, stopped answering.

After World War II even the file relating to the period 1886–1910, when the cube was at the museum, had disappeared.

This is certainly strange. It is even stranger that

there are several hundred stories like this. *Scientific American* is full of them.

Here is one of them, fully reported in the early days of that important magazine (volume 7, page 298, June 1851). According to the account in the magazine, a metallic bell-shaped object measuring four and a half inches in height, six and a half inches in base width, two and a half inches at the top, and having a thickness of one-eighth of an inch was uncovered when a solid rock was dynamited. The object was made of a metal resembling zinc; from the description, it sounds like a silver alloy. An inquiry concluded that the object was of considerable antiquity: the rock that had been dynamited was itself several million years old.

The object circulated from museum to museum and then disappeared. It was not found again.

There are too many descriptions of this sort of object for us to deny that manufactured, metallic objects have been found in very ancient rocks as well as in coal veins. We can also stress the fact that these objects have strangely disappeared.

According to the previous chapter's definitions, we should regard the presence of these objects as a working hypothesis, their mysterious disappearance simply as a conversational hypothesis; for it is well known that museums are in the habit of "burying" objects that do not seem to them to coincide with current theories, or that are not beautiful to look at. The vaults of the Smithsonian Institution in the United States, for example, are full of crates of incomprehensible objects that no one is studying. The same situation exists in other museums, including the Museum of Prehistory of Saint-Germain-en-Laye.

However, to return to the subject of real interest: at the time Dr. Gurlt discovered his cube, no one thought it possible to record data in the microcrystalline molecules of a magnetic alloy of which such cubes were made. Yet that does not mean such records could not have been stored in the cubes. Other objects of this sort are no doubt to be found in nature if we were but to look. Their owners can without a doubt retrieve them at great distance by means of a magnetometer; for the objects, when they receive a certain signal, must be able to indicate their exact position through an answering signal transmitted by magnetic resonance.

Nor need we confine our interest to magnetic recording. For example, Carson Laboratories of Bristol, Connecticut, has succeeded in photographically reducing the image of a magazine page eighty-five thousand times, in housing this image in a crystal, and then in retrieving it. Other researchers are trying to achieve registrations in crystals by successive layers, like the superimposed pages of a book. The registration of a hundred thousand average-sized books in a crystal the size of a sugar cube is currently being discussed. And it is not at all out of the question that certain precious stones may be registrations that are designed to be retrieved someday or that have already been submitted several times to "information retrieval."

It would be interesting to know whether or not

satellites have registered signals of unknown origin coming *from* the earth. Recently, some satellites in the Explorer series detected a radiation of this type: it came from earth; efforts to explain it as the result of a natural phenomenon have proved unsuccessful; it did not come from any known artificial source—radio, television, radar, etc.

As a matter of fact, this radiation was similar to that emitted by Jupiter's red spot. But since the earth has no red spot, it is difficult to apply such an explanation, which is purely theoretical anyway, to earth radiation. Jupiter's radiation is modulated (one could attribute this modulation to the act of intelligent beings, but one could also attribute it to the satellite Io and its periodic interruption in passage of Jupiter's radiation), and earth radiation does not appear to be modulated.

In short, we do not know if the unknown signal comes from the earth, or if it is emitted by the upper atmosphere or by belts of charged particles that surround the globe. Until another theory is established, we are free to think that it is coming from a recorder signalling its position.

Likewise, we may think that satellites equipped with detectors other than those that operate solely on the radio principle will uncover other radiations emitted by the recorders. Detectors in orbit around the earth, capable of registering gravitational waves, neutrinos, and eventually, tachyons, will probably show that recorders are operating all over the globe. Perhaps we will succeed in locating them through this means.

Likewise, it is probable that a certain number of

recording devices are in orbit around the earth. These recorders intercept a radio message and retransmit it with a certain delay to an unknown destination.

This idea, which I first advanced in a book entitled *Listening in on the Planets*, is now widely accepted. As eminent a scientist as Professor Roland Bracewell, director of scientific research in the area of radio technology for the Australian government, along with numerous other scientists, has been picking up abnormally retarded echoes in radio transmissions since 1929 and in television broadcasts since 1950. Television broadcasts have been received from stations that have not been operating for three or four years. And like echoes in time, radio broadcasts have been received several days after they were transmitted. This phenomenon has been observed since 1930 by Sturmer and Van Der Pol. Bracewell has made a detailed study of it. He thinks that automatic vehicles, similar to our satellites, are recording our signals and our radio broadcasts and retransmitting them to an unknown destination when the conditions are favorable.

No other scientific explanation of these delayed echoes stands up. There is no object in space on which waves could be reflected and return several minutes, several months, or a year later. Nothing in our knowledge of the structure of time can allow us to think that time is holding back the electromagnetic waves as in a trap and returning them. (If such a phenomenon were possible, it would explain many other things besides the delayed echoes. But that's another story.)

Having noted this strange phenomenon, we should say that the majority of the recorders are certainly on the earth. We find them accidentally in coal or uranium mines, or in quarries when dynamiting rocks. But it is equally certain that some of them have already been found on the surface and that they are shut up in crates, deep in the vaults of scientific institutions, covered over by other crates and by thick layers of dust. Obviously, objects that contradict theories obtained from archeology go down into the vaults first. If they later escape, it is only by pure chance.

It was in such a vault that an eminent nineteenth-century English scientist found a lens that had come from the ruins of Nineveh, a lens that had been machine cut. It still exists. It probably formed part of a perfected optical instrument, more perfect than the instruments of the nineteenth century and of course than those that existed during the time of Nineveh.

The lens was found by accident. And I am absolutely sure that a systematic exploration of the vaults of different museums and a methodical reexamination of objects labelled "art objects," "cult objects," and "unidentified objects" would yield numerous data and would be more profitable than a great many archeological expeditions undertaken at great expense.

Various studies on Central and South America have noted large spherical objects, sometimes three yards in diameter, placed on pedestals. There are no local legends attached to these spheres, which seem to be more ancient than man in these countries.

These objects could clearly be recorders of another type, placed on their pedestals in days of yore by some totally forgotten race. For even if it is easy to imagine a natural process that could produce a sphere, it is impossible to conceive of such a process as being able to shape a pedestal and place the sphere on top of it. The spheres must have been manufactured. But how and for what purpose? Right now, no one knows.

Or consider what happened when, in 1962, a steel object weighing about twenty-two pounds fell into a street in Manitowoc, Wisconsin. It was clearly a machine fragment. At several points on its surface, the steel had melted. The object was taken to the Smithsonian, which announced that it was a manufactured object, then fell into silence. Today, the object in question has disappeared under the dust of the museum.

In the same period, another object disappeared all by itself. It had fallen into a lake under the eyes of a fisherman, Grady Honeycutt of Harrisburg, North Carolina. According to the witness, the object resembled a football covered with metal rods like a "metal hedgehog." By the time the police arrived, the object had begun to decompose and looked then like a mass of metallic threads wound up into balls.

The following year, an analogous object was noted in Dungannon, Ireland. This one had only four metal rods—no doubt it was a poor variant of the preceding one—but it was incandescent. The Irish army retrieved it and after that it was no longer heard of.

One could obviously think that these were fragments of spy satellites; however, objects of this sort have been in the process of being gathered since the birth of humanity, well before there were spy satellites put into orbit.

Too, if these objects come from spy engines, these engines were not manufactured by men.

What is to be hoped is that the next angled object discovered will be carefully examined, especially with a mind to extracting its signals.

3

The Nasca Visitors

The gigantic figures that one can observe outlined on the ground of the Nasca plateau in Peru were not discovered until 1947, when they were spotted from the air. Only visible from above, they have been studied in complete detail by the German archeologist Maria Reiche.

The Nasca plateau is about forty miles long and six miles wide. It is covered with small iron and silica stones, coated with a black patina. These stones were, no one knows how many millennia ago, moved about to form a complex design, perfectly visible from above and completely impossible to spot from the ground.

The design is made up of straight lines and large surfaces in trapezoid form, rather resembling one of our airports as seen from an airplane. Also to be seen are giant spirals analogous to what telescopes show us of the spiral nebula. There are also large-sized nonhuman figures, or figures of gods or extraterrestrial beings: take your pick.

We have absolutely no idea how, working from

the ground, anyone could execute figures like these in such perfect proportion. On the other hand, if the work was controlled from *above,* from a flying machine hovering in one spot, then there is no problem. Of course, it is not likely that archeologists will accept this hypothesis for some time.

Studies carried out on the Nasca complex show many interesting features. For example, marked out from a central square measuring nine feet on a side, are twenty-three straight lines, some six hundred feet long. It has been proved that two of these lines are astronomically oriented: one to a point in the sky corresponding to the solstice, one to the equinox. The other twenty lines are unexplained. Perhaps the whole complex is a computer like the one at Stonehenge. There is much still to be explained.

The whole thing suggests, to an open mind at least, an astroport that might also be a sacred place of homage to visitors from space, whose representation forms part of the construction of the astroport. Beings with extraordinary technical means could certainly have moved these stones with the help of machines operating on the ground. In any case, the stones were reassembled in piles, seemingly sacred; they have not been moved in millennia.

The nonhuman figures, which the inhabitants of the region could not see since they did not have air vehicles, nevertheless influenced their art, and we find them on much Peruvian pottery. But there is no reason to believe that the culture commonly known as the Nasca culture, which lasted from 300 B.C. to 400 A.D., was *responsible for* the Nasca designs. We could just as easily say that it was William the Conqueror who built Stonehenge.

There is too much of undetermined origin in Peru to know what should be linked to Nasca.

Thus, according to Peruvian legends, the demigod Manco Capac, founder of the empire, arose out of the depths of the earth through what the story calls "the splendid opening." There were two more of these leading to nether worlds. These gates were supposed to have been located in a hill called Tampu-Tocco, some twenty miles southeast of Cuzco. It goes without saying that these gates were never found, that superstitious terror prevented anyone from looking for them. Manco Capac is supposed to have been turned into stone and preserved for eternity. His petrified body has never been found, though the Spaniards of the conquistador period saw the mummies of the ten emperors who succeeded him.

It is not easy to date Manco Capac, but he seems to have lived relatively recently when compared to the unknown people who built the civilization on the Marcahuasi plateau that was discovered and studied by Daniel Ruzo. This civilization goes back to at least 10,000 B.C., if not further. The oldest civilization in the world, it has left strange monuments engraved in rock by a technique that makes images change with the seasons. Thus, a figure that looks, in winter, like an old man becomes transformed in summer into a radiant young man: perhaps the origin of the solar myth of death and resurrection that we find in so many religions.

Between 12,000 B.C., the time period of the Marcahuasi civilization, and 3500 B.C., the period in which we can probably locate Manco Capac, we can place most Peruvian mysteries, including the construction of the Nasca designs.

The dating of the fantastic ruins of Tiahuanaco and their possible relationship to Nasca has been harshly debated. Posnanski gives them a fantastic age, dating them before the emergence of the South American continent. On the other hand, local legends say that Tiahuanaco was built before the stars. And some make Tiahuanaco the center of a lost megalithic empire. Perhaps they are right.

The gigantic megaliths of Tiahuanaco are located at an altitude of twelve thousand feet, in an icy, moon-like region practically devoid of vegetation. It is extremely disconcerting to find the traces of a powerful civilization on such an inhuman site. The Tiahuanaco structures, four in number, each measure thirteen hundred fifty by three thousand feet: pyramids, pillars, and the famous "gate of the sun," which is nine feet high and twelve feet wide and was fashioned from a single block of andesite mineral weighing about ten tons. The "gate of the sun" bears inscriptions in which it is thought we can identify a Venus calendar, spaceships, and extraterrestrial beings. These interpretations are, of course, still disputed.

We also find at Tiahuanaco a statue representing a human being with a rather disturbing appearance. It is made of red stone, twenty-four feet high and three feet thick. It could be said that this full-sized statue represents a giant. Who can say?

Clearly, something extremely odd happened in Peru, just as in the Near East. The serious archeologists place the construction of Tiahuanaco between 1000 and 1300 A.D. Their arguments are neither better nor worse than those of the romantic archeologists; in fact, the classic archeologists reacted vigorously to the excesses of the romantic, who no doubt exaggerated a little.

According to classic history, the Inca emperors succeeded one another as follows: Manco Capac, died 1105; Sinchi Roca, 1105–1140; Lloque Yupanqui, 1140–1195; Mayta Capac, 1195–1230; Capac Yupanqui, 1230–1250; Inca Roca, 1250–1315; Yahuar Huaccac, 1315–1347; Inca Viracocha, 1347–1400; Pachacuti, 1400–1448; Toupac Yupanqui, 1448–1482; Huayana Capac, 1482–1529; Atahualpa and Huascar, 1529–1533.

In this chronology, we note that Manco Capac does not appear as an ordinary historical figure; he seems to have come from outside and to be unlocated in time.

According to historians, it was one of these emperors, or several, who must have built Tiahuanaco. They do not, however, provide a single piece of proof. The Incas did not use writing, but rather quipus, cords with colored knots, for transmitting information. The Spaniards could never find anyone willing to read quipus for them, and those that have come down to us remain completely indecipherable.

Of course, arguments that date Tiahuanaco from 250,000 years ago are no more convincing. Between the two extremes, we find historians and popularizers who keep their heads. These include L. and C. Sprague de Camp, who say, of the lost empire of Tiahuanaco, that these were two stages in its culture, one dating to pre-Christian times. And they point out that the imperial civilization had already declined before the arrival of the Incas, speculating that some of the Manco Capac legends and those of the pre-Inca empire were connected to real events in Tiahuanaco's history.

Certainly, in the circumstances, it is permitted, while sharing this point of view, to muse a little and imagine that, in Peru as in the Near East, a control was effected by the Intelligences. This control consisted of a certain transfer of data from the Intelligences to humanity—whence the various abnormal phenomena that we observe in both Palestine and Peru. From these, I have chosen the Nasca designs because they are obviously a signal directed at space and perhaps even a construction achieved from the sky. This is less obvious in the case of such marvels as the immense megalithic fortress of Sacsahuaman.

The later is extraordinary, nevertheless. Blocks up to twenty-five feet wide and weighing two hundred tons were fitted so exactly—and without mortar—that it is impossible to pass a knife blade between them. In spite of all the rationalizations, we have to suspect an outside technology or one coming from a higher civilization. In any event, this technology was of a much higher level than that of the Incas, who were not acquainted with either the vaulted arch or the wheel.

As in the case of the Nasca figures, although to a lesser degree, we are dealing with a technology that was far superior to the local technology and even to what the local technology would be a century or a millennium later.

This same superiority appeared in medical surgery. The pre-Inca Peruvians had an amazingly advanced knowledge of surgery, more developed than that of any other ancient people. They used the tourniquet and the forceps and employed coca-based anesthetics. They bandaged surgical incisions with gauze and absorbent cotton. Medical historian R. L. Moodie cites among the operations they practiced amputation, excisions, trepanning, bone transplants, cauterizations, and "other, less obvious processes." Operations of these types are represented with some precision on pre-Inca and pre-Columbian pottery.

We have indications of the use of hypnotism as an anesthetic before operations. All of this is so completely different from what we know of other primitive peoples that we must ask questions. In Bolivia today there is still a caste of medical priests who retain the ancient secrets: the Collahu Aya. They do not talk; they travel throughout South America treating illness and carry with them, in boxes, medicines that we have not been able to identify. In this case, too, perhaps we have a survival.

The most fantastic hypotheses obviously raise even more fantastic questions. Why did the Intelligences choose Palestine and Peru to carry out controls or interventions? I confess that it is completely impossible to answer this question at the moment. Perhaps one day more refined techniques in the study of geophysics will permit us to reconstruct the history of the radiation belts that surround the globe and to learn what gaps these belts have had in certain recent periods. Currently, there are gaps at the poles and one above Africa that apparently have absolutely no function. All of this assumes that the radiation belts can interfere with a large-scale operation on the earth, which is true given the limits of our technology, but which is absolutely not true if we imagine a control of natural forces far superior to our own.

Unfortunately, it is only possible for us to think in the framework of what we know and can do. Failure to do this leads to ridiculous predictions—the stuff of science fiction.

Therefore it is in the framework of *possibility* that I will try to describe what happened at Nasca.

There are two variants to this description:

1. The employment at Nasca of a vehicle analogous to our air-cushioned vehicles; for example, the aerotrain. This machine, which the natives must have looked at with a mixture of admiration and fright, flying two or three yards above the ground, lifted the stones by "inhaling" them, then projected them onto the sides of the drawing for the natives to arrange in piles. With this order of

technology, the Nasca designs could probably have been completed in a month. With the job done, the machine, probably teleguided, rose up into the sky and was retrieved by a spaceship or a capsule in orbit.

There are numerous legends of this sort in Peru; but the specialists, as we know, refuse to take the legends seriously.

2. A second version assumes that the designs were made by the local population following the orders, if not of the Intelligences, at least of some higher race representing them. This raises certain problems: since the designs make no sense except when viewed from above, how could the local populations understand what they should do?

Professor J. Alden Mason, a respected scientist and curator emeritus of the University of Pennsylvania, suggests that the Nasca designs could have been made from a reduced model—for example, a drawing. But who could have made the drawing? And this hypothesis does not take into account the enormous difficulty one encounters in getting primitive people to work from a plan or a photograph. This was tried, notably during World War II in the building of airports, but no one ever really succeeded. It seems to me that Professor Mason's hypothesis has to be rejected.

We can imagine a scene similar to those in the Bible and in various legends:

After a number of miracles, a voice is heard, the people assemble, and each worker receives exact instructions as to what he must do. The entire maneuver is directed from a helicopter, balloon, or some other machine capable of hovering—one that we still have not invented. The work lasts for years, even decades and, when it is finally done, the Intelligences or their representatives go away. They leave behind them a certain number of techniques, particularly the machining of platinum, which has been known in Peru from earliest times and was not mastered until 1730 by Europeans, and then only with methods inferior to those of the Peruvians.

What was all this work for? No doubt to build a sort of astroport, where the recorders collecting data about the earth could be brought for retrieval by relatively simple, optically self-guiding machines. This would be the part of the design that looks like an abstract drawing.

As for the figurative part picturing nonhuman beings and strange objects, different possibilities present themselves.

It could have been there just as a simple decorative motif; art is often mixed with technology. We ourselves plant flowers at airports.

The figurative section could also represent, according to a system of notation current at the time, the heavenly constellations with which the visitors were identified.

Finally, the interpretations might be completely wrong, and what the archeologists call an "unknown animal" might represent the visitors; what they call a "whale" might really be a spaceship. And so on.

An archeologist, even one of the present day, looking at a drawing of the LEM vehicle poised on the moon would conclude unhesitatingly that it was an insect. We would no doubt have to decipher the Peruvian "abstract" figures carefully to find out if the spirals correspond to known nebula, as is the case with similar spirals discovered in Siberia.

Up to the present time, these figures appear to be the only figures in this region. Satellite observation of Peru will certainly provide interesting information. But while no further figures have been discovered, on the Nasca River, at Cahuachi, a strange place has been identified, baptized "the wood Stonehenge." It is a collection of wood columns and forked tree trunks, averaging six feet high, that are several millennia old, their wood having been preserved through a special process. Beside it, we find tombs from the Nasca culture. The most likely, although prosaic, hypothesis is that this group of columns and forked trunks supported a vast canvas tent covering a sports arena or meeting place.

Or, since the Nascas were beheaders, as we can see from their pottery and cloth, perhaps the pillars and forks of their wood Stonehenge were decorated with enemy heads after each victory.

Or perhaps it was built for something completely beyond our imagination.

The Nasca mind seems extremely different from our own. They are, for instance the only people we know of who made their tombs bottle-shaped. And we find in their tombs pottery with pictures of feline creatures wearing diving helmets. It would be tempting, obviously, to think that these were pictures of the Nasca visitors. But the truth is undoubtedly more complicated.

The Nascas did not raise pyramids or megalithic, cyclopean structures. This is why it seems so unlikely that the so-called Nasca culture, which we can locate as existing probably between 300 B.C. and 400 A.D., was responsible for the Nasca designs. The latter must already have been in existence; man, after all, appeared in South America at least ten thousand years earlier than he did in Europe.

Radiocarbon dating does not work for South America, as Professor J. Alden Mason confirms. Skulls have been discovered in the Andes of a race that is up to now unidentified; these remains are at least twenty thousand years old and associated in tombs with mastodon bones. It is completely possible, therefore, to think that these men executed the Nasca designs or saw them executed.

We would probably achieve much greater precision if we could date the Nasca designs by analyzing the abstract portion with a computer, as was done for Stonehenge. It is true that if we obtain a date for it very much prior to that of any known civilization, the archeologists will protest vigorously, since such a fact would not jibe with their theories. They have, however, been forced to accept the paleolithic lunar calendar established by Alexander Marchak, a pebble-and-bone calendar that is at least thirty thousand years old.

The Nasca calendar—if it is one—is perhaps even older.

It is impossible to radiocarbon-date gold or platinum objects, and we find some truly singular ones in Peru, worked by methods that, as J. Alden Mason writes, were either invented in Peru or introduced from no one knows where.

But it would be nice to know. Not even the Egyptians knew about platinum. Unfortunately, the Spanish plunderers melted most of these objects and turned them into ingots, which were easier to take back to Spain.

It will probably be easier to uncover the underlying explanations of the Nasca designs, even if some details escape us, than to succeed in proving that certain techniques were spread from Nasca. We should not forget that, just like the Spaniards, the Incas were conquerors who destroyed the civilizations preceding them. As a result of this double destruction, there is little information left on the pre-Inca civilizations.

The Inca history we actually know begins quite recently, in 1438 A.D. with Emperor Pachacuti, the successor of a semi-legendary figure named Viracocha, whose son he overthrew. Pachacuti began the construction of a vast empire that extended on the north as far as the Equator, the center of which was in Chili. This empire, thousands of miles in length, covered a vast area. Pachacuti can be compared to Alexander the Great, Genghis Khan, or Napoleon, and like them he was a great destroyer—burning documents, destroying towns, razing monuments, and having others built.

His successors followed his policy of conquest and laid their hands on all of South America. Then they made war on a civilization that we still have not been able to trace but that sometimes is located on the Galápagos Islands, sometimes on the Santiago or Floreana Islands. The latter is Thor Heyerdahl's point of view, but as is sometimes the case the facts do not support his brilliant theories.

We do not know anything about this island civilization; it seems to have disappeared completely. It would be interesting to know whether or not the inhabitants were in the habit of drawing inscriptions directed to the sky.

Emperor Pachacuti abdicated at the great age of one hundred and twenty-five and set his son on the throne. Under the latter's reign the Inca empire absorbed the remains of civilizations that have not been identified precisely. What this empire might have led to before the arrival of the Spaniards is almost impossible to imagine. At the time of the Spanish invasion, the Inca empire covered an area as large as France, Italy, Switzerland, Holland, Belgium, and Luxembourg combined.

In about 1523, the Indians were sufficiently organized to attack the empire. With them they had a Spanish adventurer by the name of Alejo Garcia, who had been shipwrecked and was the first European to come face to face with the Incas. Pizarro did not come until ten years later; according to the chroniclers of the time, if Garcia had been able to write his memoirs before being killed in combat, he would have had many marvels to tell, for he had sources of information other than the Incas. His death in Paraguay was certainly one of the great missed opportunities of history.

The Incas, or at least their ruling class, seem to have had no historical curiosity and did not concern themselves with the mysteries they encountered in the conquered countries. They were posted, however, on the prophecies concerning the return of the Great Old Men, who would be accompanied by fair-skinned men and strange animals. Unfortunately for them, they took the Spaniards to be these demigods returning from the Beyond, a fact that allowed a hundred and eighty Spaniards to conquer an empire of sixteen million Incas.

Such a conquest is without parallel in history. We may see it repeated, perhaps, if Earthlings are one day taken for the ancestral gods of some other planet.

Unfortunately, the thick-headed brutes who effected the conquest of the Inca empire did not interest themselves in megalithic cities or in inscriptions aimed at the sky. Later, in 1653, a priest, Barnabas Cobo, wrote a four-volume history of the New World that is currently our main source of serious information on the Inca and pre-Inca peoples. No serious sifting of this work, with respect to fantastic realism and the interventions of extraterrestrial beings, has yet been undertaken.

On the other hand, every lunatic in creation seems to have thrown himself on the megalithic cities of Peru in the first half of the twentieth century. Luckily, Nasca escaped them, and the first serious book on the subject, Maria Reiche's, appeared in 1949.

It is regrettable that in the course of their expansion the Incas conquered the Chimus, a civilized people who lived in towns organized into districts with densely built-up streets. The Chimus showed great interest in the world in which they lived, and they had a complex society. Their language, which was called Yunga, has been lost. Their capital, laid waste by the Incas in 1500, was called Chancan; it was the size of Paris and was made up of residential structures measuring hundreds of feet broad and long. The Chimus had an organized central government, but we have not found any trace of religion.

Between 1000 and 1500 A.D., the Chimus built towns and invented techniques for manufacturing bronze and melting metals in molds. All of these inventions are relatively modern, however, and strike us as independent advances rather than rediscoveries of lost secrets.

It is unlikely that a civilization capable of building cities of this size and possessing a centralized government had no method of recording information, even if they could not write. But we know nothing of this method, or methods. The town is full of geometric bas-reliefs that we cannot interpret. One could even suggest that it was the ancestors of the Chimus, and not those of the Incas, who carried out the Nasca designs; but we cannot go any further than this for the time being.

A small number of serious archeologists believed in a megalithic empire centered around Tiahuanaco. If this empire did exist, it preceded not only the Incas but the Chimus as well. However, several points of this hypothesis are still debatable, especially with regard to the empire's technology. This is the case with the copper clamps that hold some of the megalithic blocks in place at Tiahuanaco. This is a completely original technique, one not found anywhere else in Peru. Obviously, again, although we could see it as a sign of secret techniques left behind by the Nasca visitors, we could also see it as an ingenious independent discovery.

The ground at Tiahuanaco is just beginning to be worked, and no one has yet made a sensational discovery. To my knowledge, no similar excavation has been made at Nasca. Perhaps when we do excavate it, we will find crypts in which certain objects were stored before being shipped to the sky.

It is surprising that none of the Nasca drawings can be identified with a comet. The Incas had a particularly marked fear of comets and gave up their fight against the Spaniards when a comet appeared. Might this not relate to a memory of strange phenomena connected with the sky?

On the other hand, there is certainly a resemblance between the Nasca drawings and the Venus calendar, and perhaps even the solar system.

An attempt has been made to interpret certain Peruvian bas-reliefs and ceramics as maps of the world of which Tiahuanaco must have been the center. This effort is connected with the theories that make Tiahuanaco the origin of civilization—not just those of South America, but those of all the world. A similar effort could be made with Nasca, but I doubt that it would be profitable: the Nasca phenomenon is unusual, and it does not seem to have spread.

4

The Maps of the Sea Kings

The strange story of Piri Reis's maps began in 1929, in Istanbul, then called Constantinople. A map traced on parchment was found; it was dated the month of Nuharrem, in the year 919 of the Prophet, or 1513 A.D., and was signed by Piri Ibn Haji Memmed, the full name of Admiral Piri Reis.

Piri Reis was beheaded in the year 960 of the Prophet, 1554 A.D. By origin a Greek Christian, he was the nephew of the famous pirate Kemal Reis. He took part in numerous pirate expeditions, especially under the leadership of the famous Khair Al-Dir Barbarossa, and held the high post of Kapudan, the period's equivalent to governor, of Egypt. He pillaged Aden. He also besieged Ormuz but lifted the siege after receiving a large sum from the local government. Enemies denounced him to the Sublime Porte, and he was arrested and beheaded in Cairo. The residents of Ormuz tried in vain to get back their ransom.

Piri Reis described his voyages in books and atlases. One of his maps actually seems to have been used by Christopher Columbus. As for the map discovered in 1929 in Istanbul's Seray Library, it shows both coasts of the Atlantic and supplies an extremely clear representation of America.

The map drew the attention of a leading American researcher, Arlington Mallery. Mallery proved, by calculations that were later fully confirmed, that this map had required advanced knowledge of spherical trigonometry, and that it dated from an extremely early period, a period in which the Antarctic's ice had not yet covered the area of Queen Maud Land.

Mallery's work drew the attention of Professor Charles H. Hapgood of Keene State College in Keene, New Hampshire. Professor Hapgood had already gained a reputation as the author of a book on movements in the earth, prefaced by Albert Einstein, who personally confirmed all of Hapgood's calculations.

It was Professor Hapgood who called Piri Reis's map and others similar to it "maps of the ancient sea kings." It was he who proved their considerable antiquity and showed that their drawing had probably required the use of a flying apparatus (the same one that supervised the drawing of the Nasca figures?).

Besides Professor Hapgood, we can cite the following as among the noted specialists who have been interested in this problem: Daniel L. Linehan, S.J., director of the Weston Observatory at Boston College, who confirmed Mallery's calculations with regard to Antarctica; and the French explorer Paul-Emile Victor, who did the same. Other studies are being carried on currently, though of course the following hypotheses are mine.

From the fourteenth century on, navigators had *portulans.* As their name indicates, these were maps for navigating from one port to the next. In 1889 the Norwegian explorer A. E. Nordenskjöld began to think that these maps did not date from the Middle Ages, that they were much older. He also implied that they had been copied from an original dating back to Carthage at least, or to one even older. Attempts have been made to explain the geometric lines appearing on the maps as referring to earth's magnetism and to the compass —without success.

When the Piri Reis portulan was discovered, Arlington Mallery ascertained that the maps indicated the Antarctic and North and South America with disconcerting, not to mention impossible, accuracy. Mallery also showed that Piri Reis's map was a *copy* of a map, or of a series of maps, *older than any we possess.* These maps were made in a far distant past, perhaps fifteen thousand years before the Christian Era, by a maritime people who had knowledge of the earth's curvature, and of spherical trigonometry, and who had "aircraft" (or space vehicles?).

Mallery, an officer, engineer, and mathematician, compared Piri Reis's map to a U.S. Army map used in World War II. The latter employed the principle of equidistant polar projection and placed the center of projection in Cairo, where there was an important U.S. base during the war. The resemblance to the Piri Reis map is absolutely astonishing and furnishes proof that those who made it were acquainted with spherical trigonometry and the general structure of the globe.

On the Piri Reis map, we find the Amazon, the gulf of Venezuela, South America from Bahia Blanca to Cape Horn, and finally Antarctica—which was not discovered until 1818. However, the Piri Reis portulan corresponds not only with what modern maps show also but with the profile, obtained with recent geophysical methods, of the continent as it exists beneath the ice. The conclusion to be drawn is that the original of the Piri Reis portulan was drawn *before* the ice covered the region of Queen Maud Land. This would take us back fifteen thousand years, or even further.

The resemblance is too striking for simple coincidence. Someone drew up this map in a far distant past and copies, like that of Piri Reis or of Cronteus Finaeus, made in 1531, have come down to us. On the latter map, the dimensions of the Antarctic continent correspond very closely with those on the best modern maps. When these maps were drawn, there was ice in the western part of Antarctica, but it did not cover the whole continent. Now, modern methods in geophysics have shown that, six thousand years ago, there were still temperate regions in Antarctica, particularly on the Ross Seacoast.

This then is the most recent date to which we could ascribe the originals of the portulans, but everything leads us to think they must go back at least fifteen thousand years.

A Turkish map dated 1559, belonging to Hadji Ahmed, also shows Antarctica and the Pacific coast of the United States with extreme accuracy. But it gets even better: this map shows an unknown land forming a bridge between Siberia and Alaska across the Bering Strait! Such a land passage would explain the peopling of America by Paleolithic men, who could have come on foot from Asia.

This bridge had certainly disappeared thirty thousand years ago. It is difficult to understand how an earth civilization, known or unknown, could have been aware of it. On the other hand, we can see very well how—and this is the thesis of this chapter—photographs of the earth, taken from a satellite or from a flying vehicle, translated into maps that could be understood by primitive peoples, and recopied, would give a more plausible explanation of the mystery than the hypothesis of an extremely high civilization, all trace of which has since disappeared beneath the Antarctic ice. Furthermore, the two hypotheses are in no way contradictory.

It is possible that an inspection probe from the Beyond, in the Nasca period, revealed the existence of this civilization and that contact was then made with it. Perhaps some members of this civilization were even saved and taken back? Who knows?

In any case, the first task to accomplish, one undertaken by the Hapgood team, was to establish a correlation between the Piri Reis portulan and other portulans.

The Dulcert portulan, from 1339, is the first of this type, and the others appear to be copies of it. The precision of this map with respect to Europe and the Mediterranean is completely beyond understanding. From Ireland to the Don, the portulan bears witness to a kind of information that no one could have possessed in either the fourteenth, fifteenth, or sixteenth centuries. Its execution seems to have required mathematical knowledge totally out of proportion with that of the time. All the evidence converges to indicate that it must have been recopied, and more than once, from an original going back to the far distant past.

A Renaissance map belonging to Camerio and dating from 1502 confirms this point of view and reconciles with the known portulans. It, too, seems to have been formed on a grid that employed spherical trigonometry and perhaps a computer. Quantitative verifications made on thirty-seven points of the Camerio map confirm this point of view.

A Venetian map of 1484 simultaneously used the portulan system and the medieval system of adopting the twelve winds as reference points. It also was made with a totally unlikely accuracy, given the knowledge of the period.

The same resemblances have been found on a map of unknown origin; the only thing we know about it with any certainty is that it was engraved on stone by the Chinese in the year 1137 A.D. Since we find the same grid that we find on the Piri Reis and other portulans, it would seem to have come originally from the same unknown civilization.

Hapgood's mathematical study contains too many formulas to be reproduced here. However, his conclusion is that the proof furnished by the Chinese map shows the existence in ancient times of a civilization covering the entire world, a civilization whose cartographers drew maps of the whole earth with a uniform general technical level, similar methods, the same mathematical knowledge, and probably the same instruments. For Hapgood this Chinese map settles the question of whether or not the ancient culture that penetrated Antarctica and was the source of all the occidental maps was really a culture of a planetary scale.

While agreeing completely with Hapgood, I would like to point out that a cartographic satellite could, after circling a few times, learn more than a civilization covering the entire globe, as has been amply demonstrated by the different satellites launched since 1957.

The hypothesis of an extraterrestrial intervention does not seem to me to contradict that of great civilizations that have since disappeared. I would almost say that they are the same hypothesis.

The intervention of extraterrestrial beings could very well have accelerated the development of certain civilizations that disappeared either through their own fault or through natural cataclysms.

Ten, perhaps fifteen, thousand years ago, perhaps even more, maps of the earth were made by people who had access to every region of the globe, possessed excellent technical means, and knew mathematics. Given the accuracy of these topographical readings, it does not seem to be going too far to extrapolate and say that this "someone" was familiar with photography and had at his disposal flying machines or satellites.

The Zeno map of 1830 relates to a voyage of Venetians to Greenland. Given the accuracy of the coastal outlines of Norway, Sweden, Denmark, Germany, and Scotland, the exactness of the latitudes and longitudes of a certain number of islands, we once again get the impression that we are dealing with a modern copy of a very ancient map. This subject is still being debated.

Mallery thinks that he sees on this map certain islands that do not exist today, either because they were submerged, or because they were covered over by ice coming down from Greenland. Hapgood thinks instead that the Venetians made errors in copying the ancient map. He notes that the Venetians took Constantinople in 1204, at the time of the Fourth Crusade, and speculates that they got hold of maps similar to Piri Reis's at that time, and recopied them with more or less accuracy.

On this map we once again find a grid, but one that is distorted and probably misunderstood. The Ptolemy maps, as they were reconstructed in the fifteenth century, showed Greenland not entirely covered by ice and also placed glaciers in Sweden.

Yet these glaciers no longer existed in Ptolemy's time, let alone in the fifteenth century or in our day. However, when we reconstruct the glaciers as they were ten thousand years ago, we find them on the maps of Ptolemy. Once again, it seems that these maps are of great antiquity, ten or fifteen thousand years old, and have been copied and recopied.

The Andrea Benincasa portulan, from 1508, is also very interesting. On it we find glaciers that the majority of those who studied these portulans before Hapgood took for the Baltic Sea. This is a very surprising phenomenon.

We could multiply these examples. The first general conclusion we can draw is the following: the maps we use today are covered with a grid made of parallels and meridians. In the portulans we find the grid of that very ancient map from which all of them derive. It can be mathematically proved that the degree of latitude on this map is longer than the degree of longitude: this implies a system of projection. Now that we possess this system, which has been rediscovered, we note, for example, that the latitude and longitude of islands in the Caribbean archipelago are fixed with great accuracy. It is almost certain that whoever constructed these maps had mathematical knowledge, comparable to our own, especially in spherical trigonometry.

What do these documents show? An earth more ancient than our own; for example, an earth on which the Guadalquivir Delta practically does not exist, whereas now it is several hundred square miles. Now, at the lowest estimate, it takes twenty thousand years for a river to erode and form a delta of this size.

We also find islands in the Mediterranean that are much larger than those we know. In other words, the sea has worn them away since the period, twenty or thirty thousand years ago, when these maps were drawn.

The maps indicate glaciers in Sweden, Germany, England, and Ireland that no longer exist, but whose shape we have been able to reconstruct: these glaciers take us back ten thousand years.

Above all, these maps show a temperate Antarctic, where there is no ice. Most geologists state that the Antarctic ice has existed for millions of years, since the Miocene or the Pliocene. But they do not all agree on this point, and some feel that ten thousand years ago the Antarctic benefited from a warm climate that in some regions lasted until six thousand years ago. Hapgood himself is of this opinion, and it confirms his theory on the slipping of the earth's land masses.

Certain studies made in the Antarctic seem to confirm the existence six thousand years ago of a temperate period. Some of these studies show that this temperate period, although it came to an end six thousand years ago, lasted at least twenty thousand years. Hapgood thinks that a powerful civilization existed in this period and then disappeared.

For my part, I think that the earth was visited in this same period, and that the Piri Reis portulans are a sign of this visit. I repeat, in my opinion the two hypotheses are not contradictory.

In the meantime, following my reasoning, I believe that the Nasca figures precede the original of the Piri Reis portulans and come from the same source. I believe that the zodiac in the Nasca figures and in the portulans could be the same, and that we will find this to be the case when an analysis is undertaken. I believe that following Nasca, a detailed study of the globe was made and that a general map of it was drawn.

The two questions raised by this are: When? And by whom?

When? We have seen that a minimum of ten thousand years is set by the geological facts given in the maps. This is a minimum; we could just as easily speak in terms of twenty or thirty thousand —certainly tens of thousands of years.

By whom? Hapgood and others believe in a lost maritime civilization. We have to compare this hypothesis with the suggestions of scientists who think that the Sumerians were a maritime people, their civilization built on floating, nonterrestrial cities. It is equally necessary to consider the theory of Soviet archeologists, according to whom certain mysterious peoples (who have left us tombs in which we find only two things: a bear buried vertically and a spool of thin gold thread wound around a ceramic spool) must have lived exclusively on the Volga, in cities built, probably, on immense rafts. They left these mysterious tombs for reasons that are unknown.

This hypothesis is not without merit, but I would fill it out with another hypothesis. Legends of sea kings more ancient than the Vikings persist to our day. We find traces of them in well-documented works by novelists such as Jean Ray or John Buchan; but another hypothesis, presented in my *Morning of the Magicians* and numerous of its imitators, also needs attention: the existence of one or more lost land civilizations. I am often asked why no remains of these civilizations have been found. A double answer can be made to this.

First, this civilization could have been in the extreme north (where today Soviet scientists think they have discovered the remains of a continent, hitherto totally unknown, called Arctide), or in the extreme south: Antarctica. The ghosts of H. P. Lovecraft and Erle Cox, as well as René Barjavel, who is alive and well, will rejoice when the traces of an advanced civilization in the Antarctic are discovered. It will be one more case of clairvoyance by inspired writers.

Traces of lost civilizations have actually been discovered. Here is the detailed history of the Antikithera machine. Antikithera is an island in the Greek archipelago off which, about the first century B.C., a Greek galley sank. In 1901, some divers explored this galley and retrieved an object, corroded beyond recognition by the sea, which they took to the National Museum in Athens, where it proceeded to collect dust—until about 1960, when a noted scientist from Yale University, Professor Derek de Solla Price, reconstructed the Antikithera object. He found that it was a miniature planetarium, a machine permitting calculation of planetary positions.

This machine is as accurate as any that can be made today, representing the best in mechanics, and to obtain better results would require use of a computer. Describing his studies in *Scientific American*, the professor concluded his article thus: *"It's rather frightening."* Indeed. For such a machine forces us to admit that the ancient Greeks were advanced technicians, which is completely contrary, philosophically, to their abstract mentality, and to their contempt for machines; or to recognize that *before* the ancient Greeks there must have existed a technology, lost today, but equal to ours, especially in the manufacture of special bronzes and in gear calculation.

I maintain that the lost civilizations, whether land or maritime, were watched over and perhaps aided by extraterrestrial beings, without coming to a conclusion on the problem of whether these were the same Intelligences who lit up and then extinguished the star that killed the dinosaurs—or were intermediaries between them and us, races more advanced than ours, who served the Intelligences.

I maintain that these races and these Intelligences went on watching over our planet, and still do. I maintain that one of the signs of these Intelligences is the use of mathematics, which they have a tendency to teach whenever possible.

In South America, I ascribe the fantastically accurate calendar of the Mayas to these Intelligences: the Mayan year lasted 365.2420 days, whereas the exact figure, determined by the most modern methods, is 365.2423. The Mayas were accurate to nearly one ten-thousandth of a day. They likewise determined the length of the moon's cycle to within four thousandths of a day, and such a high degree of accuracy required very advanced mathematics.

I ascribe the Nasca figures, as well as the megalithic fortresses and the cyclopean buildings of Peru, to the Intelligences. I think that we will find signs of them in the Marcahuasi bas-reliefs when those have been completely analyzed.

In Mexico, I ascribe the Cuicuilco pyramid to them. This lava-covered pyramid has been dated by accepted geological methods as being at least seven thousand years old. It resembles no other architecture in the region, and it was the object of a cult from Mexico's most ancient times. The study of this pyramid, undertaken by Byron S. Cummings, Dr. Manuel Gamio, and José Ortiz, involved excavations that turned up various objects indicating a civilization clearly more advanced than any of the other Mexican civilizations. Later—these studies date from about 1920—methods employing radioactivity led to the conclusion that the volcanic eruption that covered this pyramid with lava and in time brought about its abandonment as a sacred place dates from 200 B.C. Research is being carried on currently, and the researchers especially hope that crypts will be discovered beneath the pyramid. Perhaps they will find mummies of those who, seven thousand years ago, built it and used it as a highly perfected astronomical observatory. A concrete roadway leading to the pyramid has been found; it indicates a high level of technology. Perhaps it was traveled by vehicles: if neither the Incas nor the Aztecs had the wheel, we cannot be sure that it was the same for their predecessors.

It is possible that the Olmecs are descended from the builders of this pyramid. Discoveries about this people are occurring at an accelerated rate and perhaps we will have definite information within ten years.

Numerous Soviet investigators think that the pyramids are a representation of the zodiacal light. This light is a cloud of dust that follows the earth as it moves, like the tail of a comet, and indeed has a pyramidal shape. It is difficult to see with the naked eye, but it can be spotted with the help of instruments.

Those who drew the Piri Reis maps had such instruments, and it is completely possible that the sight of this zodiacal light, a giant luminous pyramid in the sky, gave birth to a cult and to the building of various pyramids, those of Egypt as well as the one at Cuicuilco, which seems to be the most ancient.

It is not out of the question that various geometric structures, the pyramids, the Nasca figures, and many others, may be representations of objects that exist but are visible only by satellite, such as the zodiacal light or the radiation belts that surround the globe.

If the maps of the ancient sea kings represent the earth, perhaps other inscriptions, other monuments, picture the visible and invisible geometry of the solar system.

From this point of view, it would be interesting to take a close look at this division of the circle into twelve sections that we find everywhere, including on the map of Piri Reis, and that is traditionally connected with the zodiac. The zodiac is obviously a mythology that corresponds to nothing; in fact, the earth's axis has tipped since the time of the Babylonians and the signs of the zodiac no longer correspond to physical reality.

The planet Pluto does not tally with those theoretical deductions that have permitted us to anticipate a tenth planet beyond Neptune. This planet may exist, as may two others, farther away. Now if the solar system really has twelve planets—and this is something that would have been easy enough for those beings who had been observing it for a long time from the outside to determine—it seems quite natural that this revelation would lead earthlings to divide the sky and then the circle in general into twelve parts.

The division into 360 degrees came later, for purposes of easy calculation, as research on Babylonian mathematics has shown us.

It would be interesting to reexamine the problem of the maps in light of our totally new knowledge of the solar system. In my opinion, it would be extremely important to draw a map of the solar system with the earth's *three* moons—the two others are clouds of dust, anticipated theoretically at first and then observed and photographed—the radiation belts, various planet satellites, and the solar wind, and then to see whether or not we find any correspondence between such a map and the different solar system maps with which we are familiar. Certainly, if we could establish a solid correlation between the invisible, but now known, structure of the solar system and an ancient map, we would then have proof of outside contact or of the existence of an advanced ancient civilization.

Piri Reis actually called himself, in the notes accompanying his works, "a poor copyist" who was reproducing maps that were already ancient in the time of Alexander the Great. It is obvious that we could not expect him to know about astrophysics, since he did not even know that the earth was round. But this does not mean that those who drew the maps originally did not have this knowledge. According to recent studies, it seems certain that they were in any case familiar with the conversion of rectangular coordinates into polar coordinates.

We should note that the original Piri Reis maps contained constellations. Thus, in Antarctica we find in the place where Queen Maud Land is represented an indication of the serpent constellation, which is only visible in the southern hemisphere in the 70°/72° latitude of Queen Maud Land. And near the Argentine coast, the map indicates the constellation Argo. In the center of Brazil, we find the constellation of Taurus; in the south of Brazil, a wolf, which raises the question of whether it is the sign of a constellation or something else.

Study of the relationship between the sky and the Piri Reis map should be carried on, but unfortunately the scientist who had been primarily concerned with it, Archibald T. Robertson of Boston, died recently. It would also be interesting to study Piri Reis's numerous poems, to see if they might contain coded messages.

Generally speaking, the Piri Reis story has only just begun and a great many other coincidences will have to be examined. For example, it can hardly have been just by chance that the main Piri Reis historian in the nineteenth century was von Hammer, who was also the historian of the Order of Assassins; for the Order of Assassins had always claimed to possess accurate data on the exact structure of the earth and on unknown lands.

Piri Reis may have been heir to a tradition different from those that are admitted into the history books. A correlation between von Hammer's books and what we know in the twentieth century about invisible history should be established.

Be that as it may, it is necessary to determine how information dating back beyond a certain period in time has been transmitted in classic history. The destruction of libraries and of printed material has been much greater than has been thought generally. The Romans, in destroying Carthage in 146 B.C., burned a library of half a million volumes. Successive destructions annihilated the Alexandrian Library, the last and definitive destruction being by the Arabs after their conquest of Egypt in the seventh century A.D.

In Russia, Tsar Ivan the Terrible's enormous library disappeared without a trace.

In any case, we can estimate that less than 5 percent of the documents and treaties, etc., from more than three thousand years have survived and remain generally available to everyone.

Some documents are still to be found. The history of the Dead Sea Scrolls proves this. Nevertheless, as a whole, the ancient tradition is lost. That is why it is worth our while to analyze in depth those documents that we can trust and the monuments that do exist. It is furthermore really an advantage to limit ourselves to this kind of data, for it is too easy to be duped by shallow, not to mention crazy, commentators or simply by people who confuse science fiction and popular science.

For example, a good half of the commentators on Soviet science confuse the science-fiction stories and popular-science articles that, in the U.S.S.R., appear in the same periodicals. Then they go on to present a fantasy as some great new scientific discovery.

Incidents of this sort have occurred even in the United States, where the popular-science magazines never publish science fiction.

Thus when we do have a collection of data as rich as the portulans or the Nasca designs, it seems to me that in-depth analysis, like that envisaged for the signals from space, is called for.

5

The Baalbek Terrace

The mysterious stone blocks at Baalbek in Lebanon are enormous rough-hewn pieces of rock, sometimes reaching sixty yards in length and weighing as much as a thousand tons. These rocks had been raised as much as twenty feet from their source. One individual block that was not disengaged completely is still at the bottom of the quarry: it is over sixty feet long, twelve feet high, and twelve feet wide. To move it to the spot where the others were found would have taken the combined efforts of forty thousand men.

These facts are well established and have been since 1896. But humanity's collective unconscious has vaults that are as unexplored as those in the Smithsonian, and no one bothered very much about the problem. At most, a few "initiates," sipping absinthes in the cafes of Paris at the end of the nineteenth century, declared quite seriously that the "Masters" had put these blocks in place solely by will power.

The Baalbek affair was suddenly revived when I

translated and arranged to have published the work of a Russian professor, M. Agrest, in French. The latter declares that the Baalbek terrace was the takeoff point for interplanetary or interstellar spaceships propelled by nuclear energy. The blocks probably served as biological shields to protect the civilian population against radiation emitted on takeoff. These spaceships, Agrest claims, began their journey from an extraterrestrial base, explored the solar system, then returned to rejoin a more important craft waiting at the system's outer limits.

This theory caused a lively sensation throughout the world, and the debate continues. The last time I corresponded with Agrest, in 1968, he told me that he was setting out on a "special mission" and that there would be news. We will be hearing from him. In the meantime, it is certain that his thesis deserves examination.

First, it is clear that the blocks bear traces of having been worked with stone saws, and it has been thought that these saw traces could be used to make a definitive objection to Agrest's hypothesis. But it is not nearly that simple: blocks that were abandoned after having been cut out by a laser could very easily have been reworked by primitive methods during a later—the Roman, for example—period. We could, furthermore, believe that it was just because an aura of religious fear surrounded them that these blocks were reused in the construction of temples.

After the fall of the Roman empire, after the departure of the Christians, Arabs ascribed the building of Baalbek to djins called up by King Solomon. This legend, which certainly dates from after the building and rebuilding of the Jerusalem temple, is useless, but other facts are known. For example, there is the name Baalbek, which means the city of Baal.

The original temple, which may have preceded the blocks, was not a temple to Baal. It was dedicated to Hadad, the Syrian god of lightning, thunder, and earthquakes.

Then there are the slabs. Some archeologists claim it was the Romans who cut these slabs in order to make their construction appropriately solid, since the region was subject to earthquakes. Others, less stodgy, say that this tradition of regional earthquakes is based on the memory of other explosions, perhaps nuclear.

In any case, three slabs cover the underground portion of a temple. But the Romans were good enough engineers not to weaken a building by excavating huge underground passages beneath it—a sure way to tempt disaster in event of an earthquake. That is why it is difficult to accept the explanation that they were cut to provide solidity for a temple foundation.

Regarding the fourth slab: we have no idea why it was abandoned. The "official" version says that the Romans realized at the last minute that they would not be able to move it. This is clearly nonsense. Why would they have waited twenty years, the minimum time it could have taken to cut this slab, before realizing that they could not move it —especially when one remembers that for the temple of Jupiter at Baalbek, which they called Heliopolis, the Romans brought fifty-four granite columns all the way from Aswan. These columns were floated down the Nile on rafts, and to cross the Lebanese mountains they were housed in cylindrical wooden cradles that could roll—all of which just proves that the Romans knew perfectly well what they could and couldn't move.

Agrest's hypothesis seems more plausible: the slab was abruptly abandoned in process by the extraterrestrial beings, who were forced to leave for some reason, astronomical or other.

Concerning a Baalbek-Constantinople connection, we note that in 673 A.D., an architect, Kallinikos, fleeing Baalbek, arrived at Constantinople with the secret formula for a terrible new weapon: Greek fire. He gave the formula to Emperor Constantine IV. This weapon, a viscous product said to have burned in water, has never been reproduced, even by modern napalm specialists. The Greeks used it to defeat the Arabs in 674 and in 716, and then the Russians in 941 and in 1043. It was a devastating weapon: in the battle of 716, for example, eight hundred Arab warships were totally destroyed. The Greeks spread the tale that the weapon's secret had been revealed to Emperor Constantine I by an angel. It was not until our day that the story of Kallinikos's journey was learned. And Kallinikos was not an alchemist; he was an architect who *had been excavating*. The secret of the slabs is not the only Baalbek secret, apparently.

Although Emperor Antoninus Pius (138–161

A.D.) ordered the ancient temple of Jupiter replaced by a new temple containing the three giant flagstones, which architectural specialists have baptized trilithons, no document of that period indicates with any certainty that the stones were cut at that time, formed part of the ancient temple, or dated from a more remote antiquity. To my knowledge, no modern dating method, either thermoluminescence or paleomagnetism, has been used on them.

Professor François Bordes, who is known to readers of science fiction under the pseudonym Francis Carsac, assures me that modern methods of investigation have permitted us to discover tent-pin holes dating back twenty thousand years. I certainly bow to the accuracy of such methods, and I would like to see them used to date the Baalbek flagstones. Until that has been done, I feel that Agrest's hypothesis of extraterrestrial beings should be retained.

It would be very nice to know more about the Baalbek treasure, made up in particular, according to books from that period, of a number of sacred black stones. Generally, sacred black stones have been meteorites, like Mecca's famous Kaaba stone. Mohammed, who destroyed a large number of idols, spared this stone, which had been venerated since time immemorial. In Mexico, another stone was discovered, wrapped in the ceremonial garments of a mummy. Another is still venerated in the Indies. Personally, I connect these black stones with the recorders, and I am very curious to know if they will be found on the moon.

In the final analysis, contrary to what has been said so often, nothing proves that the Baalbek flagstones were cut by the Romans or by any known race. Obviously Agrest's explanation may be naive, and it is always unwise to attribute to extraterrestrial beings, or to ancient races, either actions or motives based on a technology that we ourselves were not acquainted with until the twentieth century; but at the same time nothing proves that the Intelligences, or races serving them, used spaceships or nuclear energy. They may have had less primitive techniques. And constructions such as that employing the Baalbek flagstones may be connected with concepts that escape us totally.

In any event, it is certain that the Romans did not have lifting devices that would allow them to handle thousand-ton slabs over long distances. They undoubtedly found these flagstones in the immediate vicinity of the spot where they built the temple. Although the saw markings obviously belong to the Romans, knowing who originally worked these stones is another story.

Preservation of secrets over long periods is one of the Near East's specialties. We know the story of the Dead Sea Scrolls. The example of the electric batteries is even more striking.

The first of these batteries was discovered in 1936 at Khujut Rabu, near Bagdad. Ten other batteries were later discovered at Ctesiphon. Mention of these batteries in my *Morning of the Magicians* aroused enormous curiosity, and as a result a study was made. The results of this study show that the batteries really dated from the second century A.D., and that Bagdad goldsmiths were still using an analogous process in the twentieth century. The secret had been well guarded.

Furthermore, these goldsmiths were continuing to practice electrogilding by using current from the local supply circuit and a rectifier.

It would be interesting to reexamine other strange phenomena found in this region. We will not discuss the Bible—too much has already been written about the subject—but we will simply say that Ezekiel's wheels deserve the detailed examination they are currently getting. Their description does, in fact, make one think of a piloted flying machine.

On the other hand, we should bring up the Slavic version of the Book of Enoch. The Book of Enoch, an apocryphal book, is not considered part of the Old Testament. It did not appear in the West until the eighteenth century, but it was found earlier in the Slavic countries at different dates, the oldest being about the tenth century A.D. In this book, Enoch narrates: "I received a visit from two men of very great height, such as I have never seen on Earth. And their faces shone like the sun, and their eyes were like two burning lamps. And fire shot forth from their lips. Their raiment looked like feathers. Their feet were purple, their eyes glistened more than the snow. They called me by my name."

Enoch visits seven worlds different from our own, where he sees creatures that have crocodile heads combined with a lion's tail and feet. When he gets to the seventh world, he personally meets the creator of the worlds, who explains the formation of the earth and of the solar system to him. Some serious Soviet scientists think that this book may be the distorted story of a visit to earth by extraterrestrial beings, and of a man who returned after traveling with these beings. They stress Enoch's statement that, for him, the trip lasted only a few days, but that when he returned he found that whole centuries had passed. This is consistent with what is indicated by relativity for an interstellar trip made at a speed approaching that of light. And Enoch's book, even if it is not contemporary with the Bible, was written well before the discovery of relativity.

Calculations made by the American astrophysicist Carl Sagan show that interstellar visits could theoretically be made every one thousand years. For my part, I am fairly skeptical of these calculations because they are based on a speed limit equal to that of light, and I do not think that this limit is an invariable natural law, or that it applies necessarily to a civilization more advanced than ours.

In any case, it is not out of the question that the Book of Enoch might be describing the Nasca visitors, those who drew the maps of Piri Reis, and/or those who cut the Baalbek flagstones. When we become fully acquainted with the Dead Sea manuscripts, of which no uncensored version has yet been published, we will probably learn more about the interplanetary war between the forces of Light and the forces of Darkness mentioned by the Master of Justice.

The manuscripts seem to say that a battle was fought in the sky, beyond the moon's orbit. But in the current state of our knowledge, we lack the elements that would allow us to shape a well-formed opinion.

It also is not out of the question that manuscripts and objects left by extraterrestrial beings await us in the area's caves. Frank Drake, a U.S. scientist, believes that such caves were marked with radioactive isotopes so that they would only be discovered by an advanced civilization—one more reason to hope that the atomic bomb, which would contaminate the whole region radioactively, will never be used in the conflict in the Middle East.

In the current state of archeology, it is difficult to know what connections might have occurred between Baalbek and the lost cities of Arabia, which were almost contemporary with the first temple, about 4,000 years ago. For the time being, the Rub el Khali desert—called the Empty Quarter—remains almost totally unexplored.

According to Silaki Ali Hassan, a modern Arab scholar, there once existed a city in the desert, El Yafri, built of enormous cyclopean blocks such as those of Baalbek. No infidel would ever have seen it, and the city should not be confused with Irem, H. P. Lovecraft's doomed city, since Lovecraft died four years before Silaki Ali Hassan's first revelations were published in the United States.

H. St. John Philby, traveler and explorer of the

Arabian Desert, claims that he passed within a couple of hundred miles of El Yafri on his journey through the Empty Quarter, and that he spoke with Arabs who had seen the city. Survey of the area is forbidden, but perhaps satellites will take a photo of the lost city.

Skyscrapers belonging to a lost civilization have been discovered in the Hadhramaut desert. Several lost civilizations existed in this area of the Wadi Hadhramaut. We have had specific data about one of them at least, Arabia Felix, which grew up between the second century B.C. and the first centuries A.D., since *1969*. A temple has been found called Mahram Bilqis, which is comparable to Baalbek. In addition, a city named Timna and many other centers have also been discovered. Prehistoric sites, some of which are seventy-five thousand years old, have been found in the Hadhramaut, and in spite of the difficulty of research in a country waging full-scale civil war, we have identified more than a hundred. From them, civilizations were born that have disappeared without leaving an identifiable trace, at least according to the information we have today.

Suddenly, about 1500 B.C., we find a Semitic civilization in the area living primarily from the export of incense, which the ancient world consumed in quantity. During the funeral ceremonies for Poppaea, Nero's wife, an entire year's production of incense from Arabia Felix was burned with her remains. The supply of incense was lower than the demand, and as a result prices were high. The infant Christ received presents of incense at the same time he received gold. King Solomon's fleets, which embarked from Eziongeber with Phoenician crews, carried incense to the entire world.

About the first millennium B.C., excavations show, Arabia Felix was made up of five kingdoms: Saba, Quataban, Habhramaut, Ma'in, and Hausan. These kingdoms were governed by magician-priests called Mukkaribs. The existence of these kingdoms and of their masters is an established fact, which is rare for this region. They had a language written with a Semitic alphabet, were familiar with pottery and metallurgy, and had enormous canals and substantial dams, especially the Marib. Their inscriptions contain legends bearing on the Empty Quarter, its lost cyclopean cities, and civilizations that have since disappeared. But the inhabitants of Arabia Felix apparently did not dare to explore the region. Their civilization seems to have had few contacts with the classic Near East. The region successively had a Jewish kingdom, an Ethiopian occupation, a Persian occupation, and finally a collapse whose cause we do not know.

Excavations continue and will perhaps determine the exact reasons for the disappearance of Arabia Felix's civilization. Several thousand inscriptions have been found and are in the process of being translated. Certain specialists in this area, such as Gus W. Van Beck, believe that this collapse was due to a disease. Other investigators admit their ignorance. The five kingdoms of Arabia Felix had lived in peace among themselves, and their cities were not fortified. The cause of the collapse was therefore not war.

The majority of their temples were dedicated to the sun or to planets, and, curiously, believers who came to the temple had to cross through a pool filled with water in order to get inside. Generally speaking, water was not lacking in the area until the destruction of the Marib dam in the had the ancient religions. Marxist historians of full decline, due evidently to economic causes: Christianity consumed much less incense than had the ancient religious. Marxist historians of Arabia Felix obviously stress this aspect, which to me does not seem sufficient to explain everything. In any case, the collapse preceded the arrival of Islam in the seventh century, and the latter cannot be held responsible for it.

Maps of the area show that the five kingdoms stood fairly close to the shores of the Red Sea and the Gulf of Aden. Today, only a few nomads enter the Empty Quarter, and one can scarcely believe the tales they tell. The various civil wars in the area have for all practical purposes prevented any survey. If, according to the nomads' tales, one believes that the Empty Quarter contains great cyclopean cities, made of blocks similar to those at Baalbek, there still is no proof.

The skyscrapers that still exist in cities of the Hadhramaut give us an idea of what the cities in the five kingdoms might have been like. They have nine stories, the first three of which are fortified and have loopholes instead of windows (it was in these first stories that arms and supplies were stored). The living quarters begin on the fourth floor and have windows. The streets are very narrow. When a skyscraper collapses, they build another one just like it. Arab texts say that these skyscrapers are reduced-scale imitations of those in the lost cities of the Empty Quarter.

If one did not believe in the lost cities of the Rub el Khali, one might imagine that these skyscrapers of the Hadhramaut were responsible for giving the Arabs, always inclined to legends, the idea for the lost cities. But these legends are too persistent to admit of such an explanation. And since the Empty Quarter has never been photographed from the sky, square mile by square mile, we cannot *a priori* deny the existence of a second Baalbek.

It, like the first Baalbek, must have been built by the kingdom that preceded the kingdom of Saba, one that possessed the technological means to handle enormous stones, or one that was in contact with beings possessing these means. The Marib dam might be a survival of these techniques.

The kingdom of Saba, even according to the most adventurous chronologies, did not go back beyond 2000 B.C. The empire or social organization that preceded must go back to some 5000 B.C. Perhaps the lost cities, now legendary, were the center of this empire.

Be that as it may, it is only in Baalbek and in the Marib dam that we find examples of the manipulation of such enormous blocks. These techniques are totally lost, and today Arabia is a country of thirst. Islamic historians never spoke of the enormous dams and fantastic canals of Arabia Felix. The Romans themselves seemed not to know about them. Yet this lost civilization, a civilization before the camel, before the date palm, and, of course, before Islam, had found a way to have, in the desert, as much water as it needed and employed a technology that its successive conquerors, Ethiopians, Persians, and Arabs, were never able to imitate.

This civilization's isolation with respect to other, contemporary civilizations is not at all surprising: only Ethiopia had the capacity for undertaking maritime contacts with this region. And there is no way of determining the precise point of contact between a higher technology and a civilization. This contact might perfectly well have taken place at Baalbek, but Baalbek could just as easily have been a place of secondary contact, stemming from some original spot which could be in the Empty Quarter. A certain number of Arab texts agree in speaking of a super-skyscraper built in the first century A.D., by a king at Ghumdan, in Yemen. This skyscraper has not been found, yet a whole people, said the chronicles, pursued by nomads, could take refuge in it. This imposing edifice had twenty floors and was built of granite, porphyry, and marble. Such architecture, surprising for the first century A.D., in the middle of the Arabian desert, is singularly reminiscent of the Baalbek technology.

One of the interesting things about the civilizations of Arabia Felix is that, though they discouraged foreign visitors, they sent commercial missions on long distances, perhaps as far as China. They imported pottery, objects manufactured in bronze, and teacups from the south of Russia.

The secrets of utilizing local water resources have been almost completely discovered. They involve the complete recovery of the region's rare but torrential rains. Canals, made of baked clay to prevent leakage of the water into the ground, distributed these torrential rains into primary feeders, then secondary, then tertiary, etc. The dams contributed to this distribution system where they existed, but they did not preserve the water. The Marib dam irrigated about four thousand acres. In addition, a system of wells, connected to the canal system, furnished a small but not negligible supplementary supply of water. It is very difficult to believe that such a complex system could have been designed without knowledge of mathematics. Knowledge of this sort was equally necessary for navigation.

And yet no trace of the mathematics of Arabia Felix has come down to us, whereas we know Babylonian mathematics fairly well—which proves once again that the entirety of a civilization's science and technology can disappear.

A day will come when the local wars will stop and when the methodical exploration of the area by satellite, planes, and helicopters first, then by air-cushion vehicles, the perfect machines for the desert, can be carried out. On that day, the giant flagstones of Baalbek will undoubtedly cease being a mystery, and the proof of a connection between Baalbek and the civilizations of Arabia Felix will be a first step toward the solution of this region's mysteries.

A second step could be made with the exploration of the Empty Quarter, where we can hope to find a large city that has not been plundered completely. Baalbek was plundered, Babylon plundered even more. This was done as much by the inhabitants to build their houses as by the archeologists lifting their treasures to carry them to European and U.S. museums, so that practically nothing was left. Later, the native populations understood the value of ancient documents, as the adventure of the Dead Sea Scrolls well shows. Every new source of ancient objects is plundered the moment it is discovered.

If it were possible to find an inaccessible town that had not yet been ransacked, and whose excavation was entrusted to an international scientific commission, we perhaps would lay bare secrets beside which those of the Egyptian tombs would seem small indeed.

A few inscriptions of the sort of the Rosetta Stone would perhaps permit us to solve the multiple secrets of Baalbek; for it is not just a matter of Baalbek. It involves a meshing of higher technologies, which are in turn mingled, as happens so often, with a meshing of legends. When we consider Baalbek and Arabia Felix's hydrological system, we think of the handling of stone and of baked clay on a large scale, with no artistic end, but with technical means comparable to, and sometimes superior to, our own.

This book is as much a factual accounting as possible. However, among its readers there will certainly be some science-fiction fans who would like to know what the connection is between the mysteries we have described in this chapter and the myths created by H. P. Lovecraft and linked to the same region.

Lovecraft referred more than once to the lost city of the Rub el Khali, which he called Irem. It was, in his description, built of cyclopean blocks, and in particular had an arch on which a giant hand was sculpted. According to Lovecraft, this hand sought to grasp the famous silver key that opened the door to other universes.

Also, according to Lovecraft, it was in Arabia that Abdul el Alhazred, the mad Arab who was supposed to have made an encyclopedia of all that was evil, the *Necronomicon*, lived.

Much of this relates so directly to the mysteries we have just described that today there are still people who go to the Bibliothèque Nationale or to the British Museum and ask for the *Necronomicon!*

Let's try to separate truth from fiction. Lovecraft himself wrote me in 1935, and confirmed to many other correspondents as well, that he had invented the *Necronomicon* in every respect. With regard to the lost city, things are more complicated. Lovecraft's collaborator on his book on the Silver Key, E. Hoffman Price, was one of the great Orientalists, knowing Islam as well as anyone and reading every variant of Arabic. He certainly supplied Lovecraft with extremely solid documentation, so it is not out of the question that the city of Irem exists, as well as the cyclopean arch and the giant hand sculpted on it.

It is not impossible that at least a part of Lovecraft's myth may be verified when the Empty Quarter is opened to exploration.

6

Visitors of the Middle Ages

Facius Cardan, father of the mathematician Jerome Cardan, recounted an adventure he experienced on August 13, 1491, thus:

> When I had completed my customary rituals, about the twentieth hour, seven men appeared to me dressed in silken garments resembling togas, with glittering boots. They also sported armor and beneath this armor one could see purple undergarments of an extraordinary splendor and beauty. Two of them seemed to be of a more noble rank than the others. The one with the most commanding air had a face that was dark red in color. They said they were forty years old, but none of them seemed more than thirty. I asked who they were, and they replied that they were a kind of men, made of air and subject like ourselves to birth and death. Their life-span was longer than our own and could extend as

long as three centuries. When questioned on the immortality of the soul, they replied that nothing survived. When asked why they did not reveal to men the treasures of their knowledge, they replied that a special law imposed the heaviest penalties on them in the event that they did this. They stayed with my father for three hours. The one who seemed to be their chief denied that God had made the world for all eternity. On the contrary, he added, the world was created at each instant, so that if God were to become discouraged, the world would perish immediately.

Facius Cardan's visitors seem to have been the last of a series who appeared throughout the Middle Ages. They were characterized by the fact that they could be communicated with, that they did not in any way claim to be angels, and that they did not bring any revelation. On the contrary, their attitude is rather like our modern rationalism. Facius Cardan's visitors denied even the existence of an immortal soul and maintained a sort of continuous-creation theory of the universe.

The alchemists and mystics of the Middle Ages quite obviously sought to connect these visitors with the various spirits spoken of in the Bible and the Kabbala, but this was certainly a case of mythological elaboration. Actually, there apparently were contacts with beings who were manufactured—"made from air"—according to Cardan's visitors.

The visitors stressed the punishments they

would incur if they divulged certain secrets. This entire tradition was to last until the eighteenth century, when, as we shall see, certain secrets were divulged.

In other regions, comparable beings were noted later than in Europe: at the end of the eighteenth century for Japan and for the North American Indians. In this period, the California Indians describe luminous humanoid beings who paralyzed people with the aid of a small tube. The Indian legend held that the people who were paralyzed had the impression of having been bombarded with cactus needles.

In Scotland and in Ireland such apparitions have been mentioned from time immemorial until the nineteenth century and sometimes the twentieth century. In the nineteenth century, we find the trace of a strange character called Springheel Jack, who was luminous at night, was able to jump or fly, and tried to enter into contact with men. His first appearances date from November, 1837 (this is the one for which the testimony is surest and most precise), and February 20, 1838; the last took place in 1877. This time the strange visitor was unwise enough to make his appearance near the Aldershot drill ground. Two sentries drew on him, and the visitor replied with a blast of blue flame that gave off an odor of ozone. The two sentries lost consciousness, and the apparition was not seen again.

These nineteenth-century cases may just be those that survived. The density of such occurrences in this period was actually much lower than in the Middle Ages, when practically every year appearances of luminous strangers were recorded. In all of the stories these strangers are inseparable from the idea of fire (the concept of energy was not yet invented). Yet when they were questioned, they invariably replied that they were neither salamanders nor fire creatures, but humans of another species.

It is tempting to ascribe to them that strange series of fires during the great plague in London; these fires suddenly destroyed all the houses that had been contaminated, and only those houses, thus preventing the epidemic from spreading and destroying the entire population of England—an interesting case of benevolent intervention.

It is likewise striking that these visitors are associated not only with fire but also with powers more or less connected with fire, in particular the power to transmute metals.

The entire Middle Ages are shot through with legends, and even solid beliefs, concerning the possibility of signing pacts with these visitors.

The rationalist conception of the Middle Ages as a period of "darkness"—ignorance—is a caricature that we must rid ourselves of. The Middle Ages were a period of rapid progress, more rapid than ours perhaps, but tending in a different direction. We have lost the ability to think the way medieval people thought, but we need temporarily to adopt the mental attitude of a man living about 1000 or 1200 A.D. in order to understand his reaction to the visitors, whom he considered a *normal* part of the world he lived in. It should be noted that the medieval men who believed in these visitors were basically men with rationalist minds, connected neither with sorcery nor with the Inquisition, which were completely different—and later—phenomena. It is not out of the question that contacts took place and that information was exchanged between these visitors and men like Roger Bacon, Jerome Cardan, and Leonardo da Vinci. In any case, the Middle Ages assumed, practically without debate, that it was possible to make contact with creatures, dressed in shining armor, whom they called demons. (The term *demon* did not then carry the pejorative connotations of diabolic evil that it does now. It was used rather in the sense of Socrates' daemon, who argued with him and furnished him ideas.)

Jerome Cardan, who seems to have reflected a good deal on the existence of demons and even stated that he encountered them, wrote on the subject: "Just as the intelligence of a man is greater than a dog's, in the same way that of a demon is greater than a man's." John Dee described their language and their alphabet in detail. In 1562, in order to study their language, he borrowed an unpublished manuscript on cryptography by Trithème, and he spent ten days recopying this manuscript by hand.

The sort of demons that are described in the Middle Ages do not offer pacts for power and have nothing to do with God or the devil. They seem most interested in the progress that men are making in natural philosophy and encourage the idea that it is possible to learn the secrets of the universe through experimentation. The phenomenon of these visits seems thus to have no connection with sorcery, still less with diabolic magic.

The Middle Ages scarcely asked itself questions about where these demons came from. Some had them coming from an unknown country on the earth and others advanced ideas very close to those we call "parallel universes" in science-fiction terminology—in other words, unknown universes coexisting with our own. As unlikely as it may seem, we find this hypothesis in medieval texts long before mathematicians were to speak of a fourth dimension. And finally, on several occasions, we find the hypothesis of interplanetary visitors expressed.

C. S. Lewis, in the notes of his trilogy—*Out of the Silent Planet*, *Perelandra*, and *That Hideous Strength*—gives various medieval references concerning luminous beings and their connections with other planets, and there are a fairly large number of documents that might be interpreted in this way, although the work has not yet been done. We should note that, at the height of the Renaissance, Kepler considered it completely natural to have himself transported to the moon by a benevolent demon who wanted to help him with his studies. This work, the subject of his science-fiction story, "Somnium," he regarded as his fundamental work.

We have noted the interest that these demons had in natural philosophy and experimentation, which in the Middle Ages was a completely new idea. With regard to this concept of experiment, we shall quote the portrait that Roger Bacon drew of his Parisian teacher, Pierre de Mariseourt, who claimed to have encountered demons:

> He is a recluse who dreads crowds and discussions and avoids fame; he has a horror of verbal disputes and a great aversion to metaphysics; while everyone is brilliantly debating the universal, he spends his life in his laboratory, melting metals, handling bodies, inventing useful instruments of war, agriculture, and the trades. He is not ignorant, however, and can read Greek, Arabic, Hebrew, and Chaldean; he practices alchemy and medicine; and he has learned to make as much use of his hands as of his intelligence.

It is this new mentality that the demons seek to study. It is with people of this sort, and uniquely with them, that they seek to establish contact. It is not a matter of a pact but visibly of a study mission. They never speak of the processes of sorcery in their visits.

During the Middle Ages, appearances of creatures with garments of light were most common. These messengers went to meet rabbis, with whom they held lengthy discussions on the Kabbala, the powers of God, the knowledge and exploration of time, etc. They stated that they knew the guardians of the sky, but were not guardians themselves.

We also see them appearing to the saints and holy men of Islam. They are always described in the same way; their intellectual attitude is always rationalist. They talk about geometry, and about a rational understanding to which God himself is subject.

We should, of course, study all these phenomena carefully. And while we may have preconceived ideas, we should not pretend that we got these ideas from revelations by unknown teachers or from manuscripts discovered in, say, hidden Tibetan monasteries—and we should never present our preconceived ideas as principles of faith. Thus I do not claim to state my opinion on the origin and makeup of these luminous demons with absolute authority. But, in my opinion, they were investigators who were sent by beings capable of lighting and extinguishing stars and were perhaps created by these beings. I believe that their immediate place of origin may have been on the earth itself, but in a region hardly localizable on a flat or a global map.

How long have these demons been pursuing their inquiry? For a very long time, certainly well before the time of Christ. To Gnostics like Irenaeus and to the Kabbalists, the messengers of God always presented three attributes: a double face, a luminous garment, and the crown of the king of glory.

This last is connected to what the Old Testament calls the Glory of the Lord, the radiance that shone around the Ark of the Covenant and that the uninitiated did not have the right to see. In the minds of initiates, this radiance, like the luminous halo surrounding the Messenger, was connected with a source of radiance and energy of extraterrestrial origin, which Claros described thus in the third century: "There is, high above the supercelestial covering, a boundless fire, always in movement, an eternity without limits. The blessed cannot know it unless He, the Sovereign Father, when he shall deem it right in his Council, of his own will shall let them see."

There is no law against interpreting this radiance and the demons' shining garments in terms of our twentieth-century mythology, against imagining that the "double face" is a space helmet, that the "luminous garment" is a force barrier producing luminous radiation through fluorescence or excitation. But we must not forget that in so doing we are replacing one mythology by another. It is perhaps wiser to content ourselves with saying that this is a new phenomenon.

The luminous demons again appeared with the first manifestations of freemasonry, as far back as the thirteenth and fourteenth centuries. It was because of these appearances that the freemasons called themselves "Sons of Light" and that they later counted years not from the birth of Christ but from a year of light that they got by adding 4,000 to the Christian year.

Interplanetary connections began to be attached to the freemasons. In 1823, Dr. George Oliver, a historian of freemasonry, wrote: "The ancient masonic tradition—and I have good reasons for believing this—says that our secret science existed before the creation of this globe and that it was widespread throughout other solar systems."

Likewise, they continued their activities during the Renaissance. They visited Cardan as well as his near-contemporary J. B. Porta (1537–1615). The latter was sole author of an encyclopedia called *Magia Naturalis*, published in 1584, in which he sought (by his own statement) to combine knowledge received from a supernatural source with experimental research. Porta was the first to study lenses scientifically, to describe a telescope, and to predict photography. His place in the history of science is therefore justly deserved. But he has been less studied in the area that interests us here.

Cardinal d'Este, who became extremely excited by Porta's works, founded an organization in 1700 that met in the cardinal's house and was called, significantly, the Academy of Secrets. Many see in it the first academy of science. For my part, I would rather see it as an intermediary between the unknown groups of the Middle Ages and early Renaissance and the Invisible College of which we have heard a good deal.

This brings to mind the Rosicrucians. According to Fulcanelli, and I believe him, among the Rosicrucians, whose writings constantly mention demons and the perpetual lamps left them by those demons,

> . . . the adepts bearing the title are only *brothers in knowledge* and in the success of their labors. No oath binds them, no statute ties them to one another, no rule other than freely accepted hermetic discipline, volun-

tarily observed, controls their free will. . . . They were and still are isolated persons, workers dispersed throughout the world. . . . Among their leaders there was never any tie other than that of scientific truth confirmed by the acquisition of the stone. If the Rosicrucians are brothers in discovery, work, and science, brothers in action and writing, it is in the manner of the philosophic concept which considers every individual a member of the same family.

I must say that I absolutely do not believe in a structured organization like the Rosicrucians, with their cells or lodges, as being the recipients of the demons' knowledge. I believe that there were meetings between free researchers, some of whom were visited by demons. Many later possessed astonishing knowledge, and we can ask ourselves where Cyrano de Bergerac got the description of a staged rocket or of wireless telegraphy.

If the demons did not publish knowledge, they perhaps carried it from one researcher to another. Perhaps they even maintained, out of range of any Inquisition, a center of knowledge where manuscripts could have been preserved. We find this sort of notion in Jewish esoterism of the Middle Ages.

These creatures of light, extremely active from the year 1000 to the year 1500, largely disappeared as time passed. In the seventeenth century, we scarcely encounter them, and not at all in the eighteenth, though Goethe once had a strange vision during a period when he was quite ill.

Yet the demons left strange objects behind them: for example, that metallic sphere of which the Templars speak in their vows. It must not only have emitted light, but light of a now-unknown intensity. In Cyprus it is supposed to have destroyed several cities and chateaux. When they threw it into the sea, a storm immediately arose, and in this region there were no longer any fish.

Perpetual lamps are found as much in the Jewish tradition of the Middle Ages as in that of Islam, or of the Rosicrucians: these lamps are supposed to have functioned indefinitely without oil—in fact, without any product that burned or was consumed. Touching these lamps was prohibited under pain of provoking an explosion capable of destroying an entire town. Several Jewish texts say that these lamps came from the watchers of the sky.

Here, too, we find the employment of forces, of energy, that seem to be physical and that do not at all correspond with the knowledge of the time. Unfortunately, none of the stories dating from the Renaissance or after that allude to lamps of this sort having been found in tombs in Germany or England can be confirmed. Some very strange and very beautiful lamps have been found at Lascaux, but we do not know how they worked.

A persistent tradition declares that the discovery of a secret tomb containing a perpetual lamp was the starting point of English freemasonry. This discovery is supposed to have occurred a few years prior to the initiation of Elias Ashmole at Warrington in 1646. There is no confirmation of it.

However, generally speaking, all attempts to link freemasonry with traditions prior to 1600 have failed up to the present. Thus, while it has been claimed that the Order of the Temple probably was not persecuted as systematically in England as in the rest of Europe, and that the survivors of the Order probably founded English masonry, bringing the Order's traditions directly into this organization about 1600 (many sincere masons believe in this tradition), nothing has been found truly to confirm this claim. We have undisputed documents that prove that masonic lodges were functioning in Scotland in 1599, but nothing before that.

All one can say is that there were contacts between masonry and "creatures of light," who came to teach. This much seems certain. But it is not tenable to deduce from this that masonry is prolonging the tradition of the "guardians of the sky."

This tradition relates to specific apparitions, beyond human control, that define a specific phase in the series of hypothetical interventions studied in this book. For a man of the Middle Ages, whether Christian, Moslem, or Jewish, it was as natural to converse with a creature of light as to receive the visit of a traveler from a far country. If these creatures inspired curiosity and sometimes covetousness for the knowledge they possessed, they never inspired fear or horror. It seems that, at a certain level of culture and above, Christians, Moslems, and Jews all quite naturally believed in the existence of a center where higher knowledge was stored and from which visitors might come to them. This is why, for example, the visit of ambassadors from Prester John's kingdom aroused curiosity but not surprise.

In the Middle Ages, the existence of this Center and of a King of the World governing from it was generally assumed, and it was quite natural that this king send messengers—just as it is natural for primitives today to see airplanes, coming from the United States or Japan, landing in regions of New Guinea or South America where there has been no contact with advanced civilization. The inhabitants of these areas know that one or more centers of civilization exist that are more advanced than theirs. But their notions of them are very vague, although some have established religions, "cargo cults," based on their perceptions.

There is so much mystery, so few clues. But we need not despair. There are some clues—for example, the famous Voynich manuscript.

Before delving into the mystery of this manuscript, however, we should say something about the art of coining secret messages, which developed in tandem with alchemy and esoterism. Just to take two examples, Trithème and Blaise de Vigenère were both simultaneously great alchemists, great magicians, *and* pioneers in cryptography. Although, thanks to them cryptography progressed to the point where it became an exact science, the art of deciphering, or making sense of a message without knowing either the code or the numbers, is much less advanced. Large computers, of course, make the work easier but they do not do it themselves. A great decipherer operates with the help of a kind of extrasensory perception that causes the information to come to him through a chaos of letters and numbers, with a kind of intuitional genius.

If this spark of genius does not strike, deciphering is not possible. A very simple idea can block a decoder completely, because he does not think; he feels the answer.

Returning to the Voynich manuscript, we can explain that it contains two hundred and four pages (twenty-eight others have been lost), and that we cannot decipher a single word of it. So why its astronomical price—hundreds of thousands of dollars—and why is so much interest centered on it?

It is because, when this manuscript was discovered in 1912 by rare-book specialist Wilfrid Voynich, he purchased at the Mondragone Jesuit school in Frascati, Italy, some authentic documents from the Society of Jesus that related to the document. And the documents were sensational. One letter from August 19, 1666, signed by Johannes Marcus Marci, rector of the University of Prague, recommended the manuscript to the attention of Father Athanasius Kircher, the most famous cryptographer of his time. Marci stated that the manuscript was by Roger Bacon. The manuscript had been presented about 1585 to Emperor Rudolph II by the alchemist and magician John Dee, who had not succeeded in deciphering it but was nevertheless persuaded that it contained the most amazing secrets. In an effort to learn its secrets, Voynich took the manuscript to the United States, where the greatest decoders, including some from the U.S. army, worked on it —unsuccessfully. Then in 1919 Vaynich submitted photocopies of the manuscript to Professor William Romaine Newbold, who was a great decoder and had rendered considerable service to the U.S. government. Newbold, a professor of philosophy, then fifty-four years old, was a man of prodigious culture. He claimed at the time that he was the only man in the world who knew exactly where to find the Holy Grail.

In April, 1921, Newbold announced his first results—fantastic. According to the texts, Roger Bacon had established that the great nebula of Andromeda was a galaxy, learned of the chromosomes and their role, and had built a microscope, a telescope, and other instruments. These revelations caused a sensation throughout the world, but many other decipherers did not agree with Newbold's solution. The latter, in any case, was only partial and covered at most a fourth of the manuscrpt, because the coding system constantly changes.

The entirety of the solution had to be found. Newbold did not have time to do it because of his death in 1926, but his work was carried on by one of his colleagues, Rolland Grubb Kent, who published some results in 1928 that were well received by certain historians, less well by others.

The major objection made to Newbold's decoding is that Roger Bacon could not, in his time, have known either about spiral nebula or about the composition of cellular structure. I am not at all in agreement with this objection: if Bacon had contacts with the beyond, he could very well have had information that seemed to come from his future and even from ours.

In 1944, Colonel William F. Friedman, who during World War II had deciphered the Japanese code, organized a multi-disciplinary group composed of mathematicians, historians, astronomers, and specialists in cryptography. This group employed sophisticated machines but did not succeed in deciphering the manuscript. However, they found the reason for this failure: the manuscript was not written in either English or Latin, but in an artificial language, invented by no one knows who (the first artificial languages date from the seventeenth century and are thus much later than Bacon), and not corresponding to any known language. Under these conditions, how could Newbold have deciphered at least part of the manuscript? By an inspired intuition, which led him to the sense across the artificial language, but which applied only to one part of the manuscript.

The investigations continue. Everyone is agreed on the fact that this manuscript makes sense and that it is neither a joke nor a hoax.

Voynich died in 1930, his wife in 1960, and his heirs sold the manuscript to a New York book merchant, Hans P. Kraus, who is currently asking a huge sum for it. Recently Kraus declared that even this figure was not enough and that, deciphered, the manuscript would be worth millions of dollars.

Decoding methods based on the "language of luminous demons," which John Dee described with a certain accuracy, have of course been proposed. These attempts have failed. One of the goals of INFO, the organization carrying on Charles Fort's work, is to decipher the Voynich manuscript. Up to now, they have not succeeded. The secret of the demons and perhaps others still more extraordinary remain in these pages covered with medieval writing.

7

Sir Henry Cavendish's Mask

If there are strangers among us, nonhuman beings who try somehow to be taken as human, they must certainly behave like the man who called himself Sir Henry Cavendish. Sir Henry claimed to be descended from a great Anglo-Norman family and was born in Nice, on October 10, 1731, under bizarre circumstances. No one knows why an attempt was made to hush up this birth, but child substitution and even stranger things have been suggested. Some of Cavendish's contemporaries even thought that the efforts to preserve secrecy were made to hide the fact that such a great English scientist was born outside his country—an amazing explanation and probably completely unlikely.

In any case, born in Nice in 1731, Cavendish died in Clapham on February 24, 1810. Although he had a poor childhood, he left, after a life filled with acts of generosity, a huge fortune. No one knows where this fortune came from: there were contradictory rumors, followed by denials, about inheritances he may have received. But one thing is certain—we have written documents on the subject—it was not thanks to his bank that this fortune multiplied. In fact, his bank wrote Cavendish advising him to invest the enormous sums he had. Bankers do not like to let money lie unused. Cavendish replied—we have this letter—by asking his banker to look after the money that he had on deposit and never to bother him again. He added that if it inconvenienced the bank to keep so much money, he was ready to take it out. He ended by saying: "One last warning: if you bother me again, I will take out all of my money."

He found excellent ways to dispense it, if not to make it multiply: every time somebody brought him a list of charitable contributions, he made out a draft equal to the highest sum appearing on it. A student whom he had employed to catalogue his library had some financial problems; Cavendish immediately sent him a draft for ten thousand pounds sterling. He kept this up throughout his entire life and still left millions, plus ownership of a canal, buildings, and other things. He was really the bottomless purse of the fairy tales. That he was an alchemist is obviously not just a coincidence.

He was brought to England shortly after his birth, and he later was a student at Cambridge until February 23, 1753. Ironically Cavendish, one of the greatest scientists of his time, did not receive a degree. No one knows why. Some have thought it was because, in this period, a degree candidate at Cambridge had to declare that he was a believing Christian and a practicing member of the Church of England, and Cavendish later stated on several occasions that he could never understand what made up religion. But this explanation, like all the others concerning him, is very weak.

Much more amazing is the fact that this man, who had no degrees, and who at the time had not published a single scientific work, was admitted to the Royal Academy of Science as early as 1760, at the age of twenty-nine. This early admittance is completely unprecedented, and, to my knowledge, has never been equaled.

What reason can be given for it? The reputation he gained at Cambridge for having prodigious knowledge? It is possible, but unlikely. Given the way gossip flies through scientific circles, the election of someone without any degrees, who has published nothing, just because he is amazingly intelligent, sounds like a highly improbable occurrence.

Yet the story has just begun. By 1773, twenty years after leaving Cambridge, Cavendish was fabulously rich—from an unknown source. He bought several houses and finally settled down in the neighborhood of Clapham Common, where the street he lived on today carries his name. Then this forty-two-year-old man began to act toward his fellow human beings with an indifference that was surprising, to say the least. He hated to have anyone say so much as a word to him. If someone he did not know dared to do so, he would bow without replying, turn on his heel, order his carriage, and immediately return home. He was incapable of carrying on a normal conversation.

He thought of women as another species, a species with which he would have nothing to do. He had a stairway built behind his house, and his female staff had to use it. If he bumped into a domestic, he fired her immediately. Here is one among thousands of anecdotes told about him:

> One night at a Royal Society Club dinner, we noticed an extremely beautiful girl who was standing by a window in the opposite house, looking down on all the philosophers as they dined. This managed to attract everybody's attention and one by one we all got up and gathered round the window to take a closer look at this beautiful child. Cavendish, thinking that we were all looking at the moon, got up to join us and when he saw what it was that we were studying so closely, he turned indignantly and strode away with an expression of intense disgust.

Cavendish was able to overcome his terror of women, however, when it was a question of protecting one. One day, in Clapham, he saw an unfortunate woman being chased across a meadow by an angry bull. He put himself between the woman and the bull, faced it, and made it take flight. Then he turned his back on the woman and went back home.

To say he was "odd" is an understatement. He wrote his housekeeper, with whom he only communicated by letter, the following note: "I want you to serve each of the gentlemen I've invited a mutton trotter. I don't know exactly how many trotters a mutton has; figure it out."

It was scarcely possible to talk to him, for he turned his back. He did have a few friends about whom we know nothing (it would be interesting to know if they resembled him). He received them in a pub, The Cat and the Bagpipe, which is now destroyed and about which we have no information. For thirty years he pursued a secret life, the details of which are completely unknown to us. He wore a violet garment, completely faded, and a bob-wig that went back to the style of the seventeenth century. He tried as much as possible to hide his face and took mysterious trips to the country in a carriage equipped with a counter of his own invention that reminds one of the meter in a modern taxi.

Then one day he rang his bell and told the servant who answered: "Listen carefully to what I have to tell you. I'm going to die. When I'm dead, but not before, go tell Lord George Cavendish." A half-hour later, he rang the servant again to tell him: "I'm not sure you understood. Repeat what I told you a half-hour ago." The servant repeated it and mumbled something about the last sacrament. "I don't know what you mean," replied Cavendish. "Bring me some lavender water and come back when I am dead."

His heirs, who had not seen him for a long time, examined his papers and discovered then that he had been a principal stockholder of the

Bank of England—not bad for someone who had not earned a penny and gave away money generously. There was also a will leaving his fortune to his family. The will required that the vault he was buried in be walled up immediately and that there be no inscription to indicate his tomb.

This was done on March 12, 1810, in Derby Cathedral. No examination or autopsy of the corpse was ever made. We do not have a single portrait of him, and we don't know exactly what work he devoted himself to in his various laboratories. The main part of his publications were not printed until 1921, more than a hundred years after his death. Even today, several trunks full of papers written by hand and instruments whose use we cannot comprehend still remain mysteries.

However, what we know for certain is extraordinary enough. The man who called himself Henry Cavendish constantly used the symbols of alchemy to designate metals as well as planets, but he performed work that was several centuries ahead of his time. Two centuries before Einstein, he calculated the deviation of light rays by the sun's mass and came up with a result numerically very close to Einstein's. He accurately determined the earth's mass and isolated rare gases from the air.

Scarcely concerning himself with publishing his work or with recognition, Cavendish carried out experiments that were amazingly original for the time. Thus, on May 27, 1775, this recluse invited seven illustrious scientists to be present at an experiment: he produced an artificial electric-ray fish and gave his guests electric shocks that were identical to those the fish normally gives. And he informed his guests—before Galvani and before Volta—that this new force would change the world. How did he know? In this same period, he found a method for measuring electric voltage by the intensity of the shock he felt on touching a circuit—which implies a rather surprising physical makeup. He thought of electricity as a fluid—unique in an era that did not know of the electron. And he figured the earth's average density at 5.48 (the modern figure is 5.52). For someone who had only a very rudimentary laboratory his achievements are amazing.

A singularly modern man, Cavendish was interested in the ancient Hindu sciences, and in the Hindu calendar in particular. He got hold of a certain number of them which he studied numerically, and drew some parallels with Chinese science.

All of this seems to me to reveal only the surface of extremely deep, still hidden research. It has been established that he was aware of the conservation of energy, for example, and that he enunciated the principle first, long before anybody else. It is even possible that he arrived at the energy-matter equivalence that Einstein later stated.

All of this was pursued in the alchemists' language—and often using their symbols. In Cavendish there was a very strange meeting of the past and the future. And yet he was as indifferent to the world around him as he was to the way he dressed (for twenty years, his tailor made the same outfit for him). He seemed never to have expressed the least interest in such monumental issues as the French Revolution or Napoleon. He was too busy with his scientific research and with things completely unknown to us.

With regard to the amazing accuracy of his research, we may note, for one example, that the discoverers of argon rendered homage to Cavendish, saying they could have only admiration for his "marvelous" research.

Indeed, in the opinion of the greatest scientists, Cavendish was an extraordinary experimenter and, at the same time, an amazing anticipator. And we cannot ignore the fact that his research plunged to the roots of alchemy.

If we add to this great achievement his indifference to human nature, his immense unexplained fortune, and his general attitude, we find ourselves faced with a person who cannot be categorized simply as an "eccentric." Furthermore, it is excessive to accuse a man of misanthropy when he never lacked charity or of misogyny when he was ready to risk his life for a woman.

Things are not so simple. It is tempting to quote Lovecraft: "Either he was born under a strange shadow, or he had found the way to go through the forbidden gate." Or another: "This face is a mask. And what it covers isn't human."

Equally, it would be easy to make Cavendish out as a unique case—if he had been the only person in the eighteenth century inexplicably to benefit from secret knowledge or considerable material help. But we know other examples of the same phenomenon. For instance, Roger Boscovitch, the Jesuit who, as early as 1756, published a treatise in which we find not only a hint of relativity and of the quantum theory but of sciences of which we still know nothing: time travel, antigravitation, and biolocation. Or Saint-Germain, the Immortal.

One is tempted to rewrite the history of the eighteenth century thus:

In the middle of the eighteenth century, an information source *X* was introduced. The event took place about 1730. *Someone* (one or more) stopped limiting himself (or themselves) to obtaining information and making inquiries and instead brought important information to Europe, notably with regard to physics and chemistry.

The introduction of this material was followed by its dissemination. This was done by messengers—it is tempting to think that Cavendish was one of these messengers—or by men of the highest ethical standards, who did not seek to draw any material profit from the knowledge they disseminated: among them Boscovitch, Saint-Germain, Benjamin Franklin, Joseph Priestley, and Count Rumford.

Later the work was taken up by men of lesser worth, parasites on the men who held this information, the most significant example being Mesmer. And finally complete charlatans appeared, exploiting incoherent fragments of knowledge, undoubtedly stolen, solely for their own profit. Cagliostro is typical of the protagonists of this last phase, which lasted until the beginning of the nineteenth century.

By 1810, it seems that all the authentic possessors of this knowledge were dead or had disappeared. Such a sudden appearance of important information was not encountered in Europe before the eighteenth century, and thereafter it was never encountered again.

In Cagliostro's case, it is visibly clear that indiscretions by sources of secret information were cynically exploited by a charlatan. He mimicked like a parrot, without understanding them, certain secrets of matter. He paraded manuscripts containing the secret doctrine, but it has been established that he actually bought these manuscripts from second-hand booksellers in London. He imitated Mesmer's hypnotism and made demonstrations with the aid of electrostatic machines that imitated Cavendish's.

Arrested in Rome by the Inquisition and threatened with torture, Cagliostro confessed completely. No one knows exactly when he died, perhaps in 1795. In 1797, when the French took the San Leo fortress where he was imprisoned and looked for him, he was no longer to be found.

It is enough to compare Cagliostro with Cavendish, who acquired limitless quantities of money without ever asking anyone for it, and who tried to remain as obscure as he possibly could, to see how much difference there is between the real thing and the imitation, between the initiate and his ape. Carlyle, the great British historian, noted quite accurately that the world is as full of semi-

Cagliostros as there are grains of sand by the sea. They are the imperfect hybrids, the failed imposters, of whom Cagliostro is the inaccessible ideal and the model example.

Thus Cagliostro represents the final phase. After him, the information from source *X* was dispersed to such a point that it became totally unusable. But in the beginning, to Cavendish alone, it revealed relativity, dynamic electricity, and atomic energy; to Boscovitch, relativity, quantum theory, ubiquity or bilocation, time travel, and parallel universes; and to the Count Saint-Germain, aluminum, on which he took a patent, and a paralyzing ultrasonic projector that the Rosicrucian Immortal called the "philosophic pistol."

All this knowledge appeared at the same moment. It was confirmed two centuries later by our most sophisticated science and our most advanced techniques.

This knowledge was connected with alchemy but went beyond it, and to designate it I would gladly suggest, by analogy, the term *alphysics*. The first bearers of this information lived apart from ordinary mortals, perhaps to hide certain physiological differences. The second series of these information disseminators (Franklin, Lavoisier) was made up of people closer to us, more human. Thus Franklin was one of his era's most prolific writers of pornography; in him we see the change from the purity and chastity of the first three.

However, Franklin possessed the secret knowledge. As he wrote to Joseph Priestley: "It's impossible to imagine the heights to which, a thousand years from now, man's powers over matter will be carried. We will learn to deprive large masses of matter of their gravitation and to give them an absolute lightness that will make them easier to carry. Agriculture will lower its work and double its production. All our diseases, including old age, will be avoided or cured. Our lives will be lengthened at will, even beyond the time they lasted in the days before the flood. And I hope that moral science will be perfected too, so that we will stop being wolves instead of men and that human beings will finally learn to practice what they now mistakenly call humanity."

This was written in *1780*—but by a man who had met Cavendish and corresponded with Boscovitch. There is a more than human knowledge speaking through him; his prediction is obviously based on knowledge of attainments that we still do not have but that we will someday achieve.

A knowledge that came from where? More or less directly from those Intelligences who are able to light and extinguish stars at will—an essentially rational knowledge, offered without request for payment and not requiring adherence to any religion—a knowledge that must have filtered over to Swift, enabling him to predict the Martian moons, and to Voltaire, who described infrared and ultraviolet in *Micromégas* and wrote to La Condamine: "Matter has perhaps a thousand other properties of which we are unaware."

It is impossible for the moment to draw up a complete list of the serious people who benefited from information coming from source *X;* there are too many of them. Some studies on this subject are under way, notably that of the American writer Murray Leinster, who is collecting all the inventions that appeared prematurely between 1750 and 1800. I myself have drawn up a list that does not completely coincide with his.

It is possible now to cite a few amazing cases. For instance, it is thoroughly established that the English mathematician Cayley invented the airplane as early as 1800. Various publications of his have been discovered—in particular, his communications to the Royal Society of Science—and a summary has been made of them by Gibbs Smith in his history of aviation. Cayley was certain—how and why?—that a motor powerful enough to propel a device heavier than air would be invented one day. Starting from this principle, he made mathematical studies and as early as 1800 described the modern airplane.

It would be interesting to know where Cayley acquired his certitude. Like Cavendish, and perhaps through Cavendish, he was in contact with James Watt, the man who invented the condenser for the steam engine. But Watt never claimed that a steam engine light enough to operate a flying apparatus could be built. Did Cayley simply start from one of Cavendish's statements? Possibly, but we do not know much. His work, too advanced for his time, was forgotten; nevertheless, we have all his publications. Was he aware of the work being carried out in Europe during his time on the hot-air engine, which if it had been pushed far enough could have led to the construction of a flying machine? No one can say today.

Le Bon, the inventor of lighting gas, was probably one of Cavendish's disciples. His work came directly out of the synthesis of water achieved by Cavendish, as well as out of his idea that it was possible to produce heat and light from inflammable gases.

Did Cayley know about the work carried out by Huyghens and Denis Papin on a piston motor employing gunpowder? This work, done at Marbourg in Germany, had been fairly widely disseminated. Cavendish, who had read everything and whose library contained every conceivable publication and was catalogued regularly, certainly knew about it. If he understood, from source *X* information, that it was possible to build a light motor for a flying machine, and to illuminate cities with gas, he could have encouraged Cayley as well as Le Bon.

Certainly the decision to divulge certain secrets was taken in the seventeenth century, in England, by an organization whose importance we are only now beginning to grasp, called the Invisible College, which included such eminent scientists as John Wilkins (1614–1672), Sir Christopher Wren (1632–1723), Thomas Sydenham (1624–1689), and Robert Boyle (1627–1691).

The Invisible College was also in contact with Isaac Newton (1642–1727) and Elias Ashmole (1617–1692). The latter had preserved in his possession the majority of the secrets of alchemy and had published a collection of alchemistic books under the title of *Theatrum Chimicum Britannicum.* The Invisible College decided, about 1660, to "reveal to the world a certain number of secrets" through an organization it created that received its charter from King Charles II of England in 1662: the Royal Society of Science. The importance of this society was immediately recognized and, as early as 1666, Colbert founded the Academy of Science in Paris.

The importance of the Invisible College is just beginning to appear to us. Its members discriminated between secrets too dangerous to reveal and those it seemed useful to publish. Therefore, the Royal Society of Science adopted the motto: *Nullius in verba,* or, "Don't believe anyone on their word."

In my opinion, as far back as 1662 the Pandora's box was opened. From that moment, all communication of secrets to be disseminated and spread has had to be made through the intermediary of scientific societies—the Royal Society of Science, the French Academy of Science, and, since its founding at the end of the eighteenth century, the New York Academy of Science. The latter organization maintains a very open mind, and it is possible to discourse seriously there on subjects that other academies will only accept with the greatest difficulty—for example, the presence of life in meteorites. If ever one were to succeed in discovering irreproachable proof of extraterrestrial interventions, one would undoubtedly be able to present it at the New York Academy of Science, whereas elsewhere it would be completely out of the question—at least on our side of the Iron Curtain. (As I have already indicated, the Soviets consider extraterrestrial intervention an antireligious propaganda argument and are ready to accept any possible evidence. Unfortunately, they accept such evidence a little too easily, and it is not always very convincing.)

Among the numerous ideas stemming from source *X* and which were clearly in advance of their time, we should cite the exploitation of rubber. No one in this era could have known that rubber would become indispensable. However, Boscovitch, like Cavendish, encouraged the Amazonian exploitation of a material that was only available in Europe in infinitely small quantities. To be so prescient as to encourage this exploitation required the rather uncommon gift of being able to predict the future, or of knowledge stemming from a higher source.

Another idea that dates from the same period found its origin in Newton: an artificial earth satellite. At the end of the eighteenth century, no doubt thanks to the combined influence of Newton's ideas and those of source *X*, the notion of an artificial earth satellite, projected into space by a cannon, rapidly took shape. We see it appearing particularly in Choderlos de Laclos, who was not only the author of *Liaisons Dangereuses* but also a ballistics specialist. In the nineteenth century, this question would be discussed in the annals of the Ecole Polytechnique, where Jules Verne would learn about it and later exploit it.

Cavendish, it seems, had a very strong feeling for the possibilities of satellites in the scientific exploration of space, but he was also interested in studying the atmosphere with the aid of balloons. On November 30, 1784, he had the first balloon ascension for a scientific end carried out by the French aeronaut Blanchard, who was accompanied by Jeffris, an English doctor originally from the American colonies. Blanchard and Jeffris carried bottles filled with water, which they emptied at carefully determined altitudes and let fill with air. Cavendish analyzed this air: it was the first study of the composition of the atmosphere as a function of altitude.

In the air the aeronauts brought back, as in the air studied at the earth's surface, there always remained "unexplainable bubbles," which Cavendish collected carefully. These bubbles were not composed of oxygen or nitrogen or any constituent of air. When they were isolated in 1895, a major error was committed in postulating *a priori* that the gases that formed them (helium, neon, crypton, xenon, and radon) were not suited to entering into chemical combinations. They were therefore called "the noble gases." We know now that this is false and that these gases are capable of forming chemical combinations, particularly with fluorine and oxygen.

For more than sixty years, researchers looking for chemical compounds of rare gases were systematically discouraged by others who explained to them that, for extremely sound theoretical reasons, such combinations were impossible. Unfortunately, the gases did not know about the theory and did indeed form combinations. We might well wonder whether Cavendish, who knew more about them than we, might not also have known that these combinations were possible.

This would explain the fact that he was able to isolate them. It would also explain why in 1921 researchers found, packed in trunks in Cavendish's laboratory, tubes filled with rare gases that he had studied by passing an electric charge through them. It does seem that the last word has not been spoken in the study of "noble" gases and that, now that research on their chemical composition is no longer forbidden, we have some amazing discoveries in store for us. For instance, it has already been established that certain rare gas compounds provide explosives much more powerful than any other chemical explosives we know of. In all likelihood, Cavendish received instructions not to make this particular aspect of his work public.

We should also note that it is a "noble" gas, argon, that is at the base of the "death ray," or chemical laser. This device, which is currently able to pierce a 4-mm armor plate, at a distance of several miles, is so simple that technicians in Cavendish's day could have built it had they had the necessary information. It appears more and more likely that Cavendish mastered this knowledge but did not reveal it.

We are realizing every day that nature holds very simple, and sometimes very dangerous, scientific secrets, but in our consideration of this "superscientific" eighteenth century, organized in concentric circles around source *X*, let us note that it had nothing in common with the occultist eighteenth century. Just as the luminously garbed demons do not figure in the sorcery trials, in the same way no information from source *X* is found in Swedenborg, Martinez de Pasqually, or Louis-Claude de Saint-Martin. We are dealing with two very different eighteenth centuries, which practically never mixed.

For example, whereas the data stemming from source *X* on the existence of extraterrestrial Intelligences are very sparse, the occultist eighteenth century swarmed with them—not just Swedenborg but people more serious in aspect abounded in information on the inhabitants of other planets. Emmanuel Kant, particularly, declared that these beings became all the richer in their spiritual life the further away they were from the sun. Thus, according to Kant, the inhabitants of Venus and Mercury had so little moral direction that they could not be held accountable for their acts. On the other hand, the inhabitants of Jupiter were in a state of moral perfection that assured them complete happiness.

It would be interesting if this theory had some connection with what we learn from astrophysics or with what we can actually observe on these planets with the help of rocket soundings; equally, it would be interesting to know where Kant got his information.

If the general hypothesis of this book is valid, it is understandable that source *X* maintained absolute secrecy over its origin. Not to be too hard on Kant, however, we should note that the idea he expressed in 1775 on the origin of the solar system remains fairly solid. Several modern doctrines, all making more of an appeal to mathematics, have repeated the same general idea.

Finally, we should pose the question of the contacts that might have existed between source *X* and the secret societies that sprang up in such great numbers at the end of the eighteenth century. Rumors concerning the existence of an extremely rich source, as much of theoretical knowledge as of practical inventions, began to spread as early as the first half of the eighteenth century. The grand master of the first masonic lodge in London, Jean Théophile Desaguliers, a man of French origin, an inventor, a scientist, and a mathematician, seems to announce the appearance of source *X*. His book on the history and doctrines of freemasonry, which came out in 1723, stresses the importance of mathematics and predicts the impending arrival of a universal knowledge brought from outside our world.

Quite conceivably Desaguliers, who taught a modern physics course at Oxford, founded a society interested in this knowledge stemming from source *X*. But alongside freemasonry, there were other, much less serious, even crazy, societies, and much work on the whole area still needs to be done.

8

Kaspar Hauser

"Kaspar Hauser was not of this world. He was brought to us, but he came from another planet, perhaps from another universe entirely." The person speaking here is no modern writer of science fiction, but someone who followed the amazing adventure very closely: Feurbach—the same Feurbach whom Marx and Engels fought so energetically, and who was a worthy opponent for them.

What could have been so extraordinary that Feurbach, who lived before Jules Verne and did not know about Edgar Allan Poe, let alone science fiction and flying saucers, would make such an amazing statement? Let's go back to the facts.

One day in May, 1828, at Nuremberg, a police officer saw a mob of what today we would call juvenile delinquents. In the middle of it, a young man, about sixteen and very badly dressed, was trying to defend himself. The officer spoke to him and the boy mimicked his words like a parrot, with no understanding. He obviously did not know that language was used for communicating.

The strange youth was taken to the police station, and when they searched him, they found two contradictory letters. One said: "Take care of my child. He has been baptized. His father was a soldier in the 6th cavalry." Examination revealed that the ink on the letter wasn't as old as it would have to be if it dated back sixteen years. Furthermore, the 6th cavalry regiment had just arrived in Nuremberg and hadn't been quartered there at the boy's presumed time of birth. The letter was obviously false.

The second was false as well. Written in an untutored hand, it said that the author, a laborer, had been employed by Kaspar Hauser. It was as suspect as the first, especially since its spelling mistakes were those a literate person might make to hide his education.

The police decided to charge the boy with vagrancy, so they could have more time to study him. Their examination showed that he scarcely knew how to walk, that if an obstacle were put in front of him, a chair, say, he would bump into it and fall. His vision was perfect, his skin white, and, to all appearances, he had never worked. The soles of his feet were as soft as a baby's. When they found him, he had been wearing women's high-heeled shoes that obviously did not fit.

He learned to talk fairly quickly, and told them that he had been imprisoned underground, that he had been fed and given toys, and that he had finally been taken from the place where he had been kept and put into a cab that deposited him in the center of Nuremberg.

An inquiry was made. The police located a fisherman who had found a bottle in 1826 containing a message calling for help and coming from a prisoner shut up in a cell on the banks of the Rhine. The message was signed Hares Sprauka. One of the police officers identified the name as an anagram for Kaspar Hauser. A search was made for the prisoner, with no result.

It was then that a first attempt was made on Kaspar Hauser's life. An unknown assailant got into his cell and struck at the left side of his chest with a dagger. The youth escaped the blow. An English nobleman, Count Stanhope, engaged a scientist, Professor Daumer, to study the case. There were a number of romantic conjectures according to which Kaspar Hauser was probably the child of a noble family. On December 14, 1833, Kaspar Hauser was murdered, probably by an unknown person who had arranged to meet him under pretext of revealing his secret.

In an era when bastards of noble families were part of popular folklore, the romantic hypotheses went very well with the general atmosphere. But one by one they were struck down. And the mystery remains.

Kaspar Hauser did not know what milk was. Professor Daumer stated that the first time Hauser came to his house, he tried to seize the flame of a candle with his hand. He had no visual depth perception; it had to be instilled in him. The place where he had been hidden was never found, and no one ever succeeded in analyzing the material on which the two false letters had been written. It was neither paper nor parchment, but from the description it undoubtedly was a kind of ultra-thin leather.

Painters made portraits of Kaspar Hauser that were distributed throughout Europe—with no results. No identification held up. But it is perhaps possible to offer an extraterrestrial explanation. In my opinion, after the period of simply auditing and recording what happened on earth, came another period, beginning a few centuries ago, in which the Intelligences began to conduct experiments. These experiments consist of introducing beings capable of arousing the most diverse reactions into our midst, and then studying the way we react—the way we study the behavior of rats in artificial labyrinths.

How does one explain the fact that the Kaspar Hauser experiment is unique? It isn't. We periodically find cases reported of people who have come from nowhere. In England, in the eighteenth century, there was a young woman who claimed to be the Princess Carabo, princess of a country that existed on no map. After a certain time passed, she made a confession that was revealed to be false. Then she disappeared and no more was heard from her.

Then there was the amnesiac they found in Paris at the beginning of the twentieth century, who had a map in his pocket of an earth that was not this earth.

And more recently there is the story of Tuared. In 1954, Japan experienced ultra-violent riots. Hoping to prove that these riots were the act of foreign agitators, the government ordered the passports of all foreigners residing in Japan to be verified. A person was found in a hotel who had an apparently irreproachable passport: no erasure and no interlineation. The photograph was exact, as well as the fingerprints. There was only one difficulty, but a large one: the passport had been issued by the country of Tuared, which did not appear anywhere on the map.

They interrogated the individual. According to him, Tuared stretched from Mauritania to the Republic of the Sudan and included a large part of Algeria as well. It was in Tuared that the true Arab Legion was organized, destined to free all the Arab peoples from oppression. He had come to Japan to buy arms.

Indignant that the existence of his country was being doubted, he gave a press conference, after which all the journalists rushed to their maps and then to their teletypes. They cabled the United Nations, the Arab League, UNESCO, everywhere: no one had ever heard of Tuared. It did not exist —not on this planet.

Before being shut up in a Japanese psychiatric hospital, the emissary from Tuared gave some interviews, in particular, to the English weekly press. He absolutely did not understand why no one believed him. His passport, on being examined, had seemed to be completely normal. It was written in the Arabic language. The only problem was that the country that had issued it didn't exist!

This person, questioned periodically by the

press, persisted in saying the same thing, and there are obviously rational explanations for his story, of one sort or another. We should recall that, in the past, they used to explain that meteorites were completely ordinary stones that had been struck by lightning. They explained ball lightning as being caused by screech owls which, having spent time in the hollow of a rotten tree, had become coated with a phosphorescent material. They did not explain, of course, how these screech owls were able to get inside the cabin of an airplane flying at hundreds of miles an hour, then explode, which ball lightning does easily, but the screech owl explanation seemed to satisfy scientific circles until 1965—in other words for two centuries.

Indeed, it is the shortsightedness of science that makes one a little skeptical of a solely psychological explanation for the Kaspar Hauser story or the story of Tuared. Apropos of psychological explanation, let me cite an anecdote.

Some fifteen years ago a convention of aeronautical engineers was held in Chicago. Some engineers from Sperry-Rand brought a gyroscope with them in a carrying case. As the gyroscope was a demonstration model, it could be activated merely by plugging it into an ordinary wall socket. After a few drinks, some practical jokers activated the gyroscope, which operated noiselessly, removed the plug from the socket and put the whole thing back in the case. Then they called a porter and told him to take the case out of the room. The porter had no trouble lifting it, but when he tried to turn and go out of the door, he found this completely impossible to do. (A gyroscope, once activated, maintains its plane of rotation and resists any attempt to modify it—which is why it can be used to stabilize rockets, ships, and planes.)

The unfortunate porter found himself fighting with a suitcase that was apparently bewitched and that absolutely refused to turn and go through the open door. After several attempts, he dropped the case, turned to the engineers, and said to them in an offended tone: "Gentlemen, you are drunk." This to me is typical of the psychological explanation applied to a physical fact.

If people like Kaspar Hauser are introduced into our civilization to arouse observable psychological phenomena, where do they come from originally? The answer to this question seems simple to me: according to the different police statistics, two million persons disappear each year without leaving any trace.

And some of these disappearances are so amazing that they immediately make one think of a paranormal explanation. The following case occurred in Great Britain, in Wales, in 1909. An eleven-year-old boy named Oliver Thomas was in the family farmhouse. Besides his parents, there were some guests: the pastor and his wife, the local veterinarian, and an auctioneer—serious people who had not drunk more than a reasonable amount enjoying an evening at home among good friends. All of these people would later supply sound evidence.

At eleven o'clock in the evening, they sent little

Oliver to get some fresh water from the well. He went out with his bucket. Ten seconds later they heard a call for help. Everyone rushed out. The pastor had the presence of mind to take a kerosene lamp with him, which lit the farmyard with a strong light: no one. But they heard, *coming from above,* the voice of little Oliver crying: "Help, they're taking me away." Then, nothing. In the snow, the track of Oliver's footsteps went from the door halfway to the well and stopped abruptly.

The nearest town was Rhayader. They went there to get the police. The latter sounded the well, searched all the houses in the area, and distributed the boy's photo everywhere. Sixty-four years later, no rational solution to this disappearance has yet been found.

A giant condor could have lifted the young boy, as in a story by Jules Verne. But no bird of this type has ever been seen in England. In 1909, there were no helicopters. Thinking that the child might have been lifted with a cord hanging from a balloon, they made a check of all balloons: none had flown over the area, or even over England, that night. No saucer, either flying or floating, no unidentified engine of any kind, was observed in England either that year, the preceding year, or the year after.

Cases of this kind are more frequent than we believe, but there is a tendency to hush them up. Nevertheless, we periodically find them, embellished with strange twists that disappear when we go back to the original documents. In Oliver

Thomas's case, there is one additional detail, but one that is very touching: for several years, in the little church where the pastor who had been present at the disappearance preached, the congregation prayed "for the deliverance of Oliver Thomas, held prisoner by men or unknown things." Let's hope that young Oliver wasn't too unhappy, or that he is not unhappy if he is still living.

We should recall in passing that it is this fear of disappearing without leaving any trace, a fear that forms one of the permanent themes of folklore, that causes and facilitates that spreading of rumors like that in Amiens, where they blame these disappearances on religious or racial minorities.

We should note too that, proportionally, these disappearances occur more frequently at sea and in the air: entire ships are made off with, or their crews, and whole airplanes vanish, in the air and on the ground. Likewise, certain areas of the sea, especially a triangle off the Bermuda Islands, are the arena for the largest number of disappearances: the frequency sometimes reaches twenty or thirty times the normal rate.

There is not any explanation for the phenomenon—at most we can imagine that somewhere, some of those who disappeared underwent a complete brainwashing and were reintroduced among us under disconcerting circumstances, by way of experiment. If we imagine that there are more perfected brainwashing techniques than our own—and this is not difficult—then we can admit the possibility of a complete transformation of the personality and an insertion of false memories designed to disconcert us. Of course we then are thrown a bone: a rational explanation that is more or less satisfying, rather more than less. And we have to admit that we aren't very curious, certainly less so than the rats in their labyrinths, which notice the experimenters and bite them when they can.

For all our lack of knowledge, a theory on a high level of reasoning, one that explains certain events as the result of experiments being performed on us, is becoming more and more widespread—for example, as defended by an extremely serious English mathematician, Erving J. Good. It is not out of the question that this theory finally will attract the attention of the scientists, and we will then see doctoral theses on the interventions of extraterrestrial beings in history being defended. Then the disappearances and the appearances will be closely studied, and the possible correspondences between them, as well as the return of certain persons who disappeared and spent centuries elsewhere but to whom only months seemed to have elapsed. The story of Enoch, discussed earlier in this book, is one that deserves such reexamination.

Traditionally, people raised by fairies return centuries later—whereas, for them, only a few months, even a few days, have elapsed. These legends have existed for millennia, in every country and on every continent. With respect to modern times, one case is particularly striking: that of the double disappearance of Jerry Irvin, a soldier. When he was on leave, on May 2, 1959, Irvin was found unconscious. He seemed to have undergone such a peculiar psychological change that it was decided to revoke his authorization to enter certain off-limits military buildings. He left the hospital and was not found again until June 19 of the same year; he was unable to say what he had done in the interval. He was examined again and confined to the psychiatric ward of a military hospital. On August 1, 1959, he disappeared from the hospital, and he was never seen again.

The Middle Ages are full of cases of this sort. The Archbishop Agobard of Lyons (779–840) made an investigation and discovered that these missing persons claimed to have visited a country they called Matagonie. Being a rationalist archbishop, he refused to believe it and even had three men and a woman, who claimed to have gone to Matagonie in an airship, stoned in his presence. What person in his right mind could believe in airships?

These unfortunate people had claimed that, for them, very little time had elapsed during the trip and that much more had elapsed for the outer world. This is noteworthy: the contradiction of time for an object moving at great speed is a perfectly established physical phenomenon. There is no reason to manifest toward this phenomenon the skepticism of the "stupid nineteenth century."

Of course, we should not expect to find all the missing persons. Some, for reasons that are incomprehensible to us, might be being kept elsewhere. The English biologist J. B. S. Haldane thought a good deal about the possibility of kidnapping by extraterrestrial beings and published a remark that always seemed extremely unsettling. He made the observation that perhaps we earthlings had capacities of which we ourselves were unaware, capacities that didn't manifest themselves on earth but that might interest *someone* other than ourselves. As an example, he cited seals, which have the ability to balance a large round ball on their nose, an ability that they scarcely have opportunity to show in their natural environment. The idea simply amuses men so much that they snatch off baby seals, teach them to balance balls on their noses, and exhibit them in circuses—without any part of this operation being explainable to the seals, whose language undoubtedly has no word meaning either ball or circus!

Perhaps we know, without being aware of it, how to do things that are completely incomprehensible and unexplainable, but that make us worth kidnapping—a disquieting conjecture.

We are convinced that we cannot do certain things, but others may know more about us than we ourselves do. Take the case of Gil Perez, a Spaniard. On October 25, 1593, Perez was on guard duty in front of the governor's palace in Manila, in the Philippines. Suddenly, he was transported to Mexico. He rushed over to the soldiers who were guarding the government palace and asked where he was. He could not believe that he was in Mexico. He told his story, which no one believed. Then he gave this proof: "Last night, His Excellency Don Gomez Perez das Marinas, governor of the Philippines, was assassinated by blows from an ax. When the news reaches here, you will be forced to judge that I am not a liar."

They immediately sent Gil Perez to the ecclesiastical authorities, for, according to the evidence, it was a case of magic. (Blessed are they who can distinguish between magic and natural laws!) At the end of two months, a ship arrived from the Philippines and brought word of the governor's death. Perez was released, but there was never, at the time or later, any explanation for this occurrence. Was it an experiment by extraterrestrial beings? Or proof of human powers that are completely unknown and normally unused?

I am sure that one day the study of these "accursed facts" will come out of the area of simple collection and enter into the area of theory. The interest of the Kaspar Hauser case, for example, comes from the fact that it has been heavily studied, although perhaps not in the best way. The studies have centered on the purely romantic hypotheses of a high-born child kidnapped in a plot against a dynasty, of a noble bastard being hidden from sight. We will doubtless have to reject once and for all this kind of hypothesis, which has never been established, to go back to the facts that we find outlined in numerous studies, and to admit the likelihood of the most fantastic hypotheses.

It is often said that the Kaspar Hauser affair is one of the great classic riddles; another is that of the *Marie-Celeste*. The seven-man crew that left New York, plus the captain, his wife, and their small child, all disappeared, at 38° 20′ latitude north and 17° 15′ longitude west, in the Azores. The missing persons took none of their possessions, not even money. And no trace of mutiny was found.

A good many volumes have offered rational and not-so-rational solutions; for example, a giant octopus carrying off the entire crew. Some of the greatest mystery and science-fiction writers, including Arthur Conan Doyle and H. G. Wells, have published stories suggesting ingenious but inconclusive analyses. A modern French writer, Yves Dartois, in a novel entitled *The Demon of Lifeless Ships,* suggests that a rare fungus developed in the boat's wood and that the spores had poisoned the crew and passengers, who then threw themselves into the sea. Dartois lists other ships, all of wood, on which a similar thing happened.

Certainly Dartois' is one of the more intelligent hypotheses, but it is not proved. Besides, entire crews have disappeared on modern metal ships. The most recent case I know of took place in 1962 in the Pacific. Others have undoubtedly occurred since then. Insofar as no absolutely convincing explanations have been furnished, backed with proof, it seems perfectly legitimate to me, as it has to Charles Fort, Eric Frank Russell, and other original thinkers, to admit the possibility that the crew of the *Marie-Celeste* was kidnapped. This obviously is not the only explanation possible, but it is one that we can logically conceive.

This is likewise the case for the crew of a

balloon used by the U.S. Army during World War II, in the fight against the German submarines. One day the balloon, which had been constantly under observation, was found empty: the three men who were in the car had disappeared. They had radios and could have signaled for help had they been threatened. But they did nothing.

The same observation applies to the recent disappearance of thirty-three U.S. soldiers whose plane crashed in the Rockies. The plane's debris was found, but there was no trace of either survivors or corpses. They are someplace certainly, but where?

It is difficult to include missing airplanes on this list: their disappearance often has a perfectly normal explanation that only becomes clear later. For instance, it is only two or three years since we learned that Amelia Earhart, the U.S. aviator who had been missing in the Pacific since shortly before World War II, was a U.S. secret service agent. There is every reason to believe that she was shot by the Japanese.

On the other hand, a very clear case can be added to our collection: the little Eskimo village of Angikuni, in the north of Canada, where *all* the inhabitants disappeared in 1930. Not only did men, women, and children totally disappear, but seven dogs tied to a tree were found dead of hunger. An Eskimo would never leave a dog to die of hunger. Even more amazing: graves had been opened and the village dead had disappeared too. Analysis of berries found in the kitchens showed that, two months before the arrival of trapper Joe

Labelle, who discovered the abandoned village, the latter had been inhabited; for those berries ripen only during a short, perfectly defined period. The Eskimos had left their guns, which is an even more convincing proof that they had not left voluntarily, for these guns represented their most precious possession.

No explanation could be found. The other Indians in the area say that the people of Angikuni were carried off by Wendigo, a Canadian forest creature they refuse to describe.

This is not the record for mass disappearance: in 1872, the U.S. steamship *Iron Mountain*, ultramodern for its time, totally disappeared. Nearly 200 feet long and 30 feet wide, it carried fifty-five passengers plus crew. It disappeared totally *on a river*, without anyone's ever finding the least trace.

Another very striking case dates from 1928: that of the Danish training ship *Kobenhaven*. Modern, and radio-equipped, it carried, in addition to a highly trained crew, fifty midshipmen of the Danish navy, all of good reputation and all good sailors. The ship itself was in perfect condition and had been carefully checked by specialists. On December 14, 1928, it left the port of Montevideo—and that was the last anyone ever saw of it.

We have less documentation on an even larger disappearance. In the course of the interminable Sino-Japanese war, on November 10, 1939, after the fall of Nanking, a regiment of three thousand men, commanded by Colonel Li Fu Sien, was sent to keep the Japanese from advancing. The regiment disappeared completely, and its radios stopped transmitting. Only a few arms and campfire traces were found. The Japanese archives, which are now available, do not mention any capture of an entire regiment at that time. And if the regiment had deserted *en masse*, the soldiers' families would certainly have known or heard about it.

This is an interesting story, but one should remember that one cannot be sure that this regiment existed at all; before Mao the Chinese army was chaotically confused. Still, if these are really the facts, the event beats all records for mass disappearance.

Whether the regiment disappeared or not, we may believe stories of disappearance in which the victims are less numerous, but where they have a reason for disappearing, a reason that touches closely on the subject of this book. For example, there is the case of J. C. Brown, a U.S. gold prospector who claimed to have found an artificial tunnel in the Cascade Mountains of California in 1904. He said that he followed this tunnel until he came to an underground chamber that had copper-covered walls. The chamber contained human skeletons, gold shields, and, on the walls, hieroglyphs that the prospector could not decipher.

Since he was not anxious to be considered a fool, Brown waited to make his fortune before talking. This took thirty years, but in 1934 he went to Stockton, California, the town closest to his tunnel, and recruited an expedition.

He had recruited twenty-four people when, on the night of June 19, 1934, he disappeared: no one ever saw him again. The police made an investigation to find out if by chance he had borrowed money against the treasure he was going to find. Their conjectures were mistaken: Brown did not owe money to anyone. There was no reason for him to disappear, unless—unless he had touched too closely on certain secrets and had to be eliminated for fear that humanity might learn about certain records much too soon.

Another miner, Tom Kenny, from Plateau Spring, did not disappear, although the discovery he made in 1936 was strange. Some twelve feet below ground, he uncovered a road paved with small square plaques a few inches on a side, of which there are no other examples in any known civilization. Later, in 1960, in Blue Lick Springs, Kentucky, a similar road was discovered, also very carefully paved. In both cases, the excavations were not pushed far enough to find out if the designs of these roads were similar to those at Nasca or other disappeared roads—indeed, even to landing strips.

A number of legends relating to vast underground domains run throughout both the Americas. According to these stories, there were at least ten million Incas when the Spaniards arrived. Forty years later, in 1571, a census taken by the Spaniards put the number of Incas at about one million. The Spaniards certainly massacred many Incas and killed even more with forced labor in the mines. But nine million? The figure seems exorbitant, and the hypothesis of an underground domain where the Incas might have taken refuge is not, *a priori*, completely unreasonable. In 1802, Alexander von Humboldt met descendants of the Incas who believed it.

Even in historical times, numerous and unexplained disappearances constitute a fact established enough to warrant our postulating the existence of Intelligences or Experimenters who capture, then experimentally release, certain human beings.

The Kaspar Hauser phenomenon is interesting enough in itself, but it seems that there have always been cases of this kind. Obviously, they were less recognizable in the past than they are today. In Paris or London during the Middle Ages, a man or woman without connections or belongings drew less attention. Today passports, visas, fingerprints, taxes, and all the other records that are kept make it possible to spot bizarre cases much more quickly.

The question we have to ask now is: "From where were Kaspar Hauser and the others brought?" They were indeed deported from someplace, but where is that someplace?

9

The Green Children

One afternoon in August, 1887, near the village of Banjos in Spain, some peasants working in a field saw two children come out of a cave, a boy and a girl whose clothes were made of a material they did not recognize and whose skin was as green as the leaves on the trees. This would be a fine beginning for a science-fiction story, but in fact this event actually took place.

Specialists from Barcelona tried in vain to identify the language they spoke and to analyze the material of their clothing. The children were remanded to a local justice of the peace, Ricardo da Calno. He tried to rub off their green color, but it was not makeup; it was the true pigmentation of their skin. The children's faces were noted to have certain Negroid traits but their eyes were of a rather Asiatic type, almond-shaped.

For five days the children were given a wide variety of foods, which they refused to eat. Finally, they were brought some freshly cooked beans, which they accepted. The boy, who was too far weakened, died. On the other hand, the girl survived. The green color of her skin gradually faded, leaving her a normal complexion for a person of the white race.

She learned a little Spanish, but when she was questioned, her statements only complicated the mystery. She described the country from which she came: a country without sun, where a permanent twilight reigned. This country was separated by a river from a bright country where the sun shone. There had been a whirlwind suddenly, accompanied by a terrible noise, and it had lifted the two children and deposited them in the cave. The young girl survived for five years before dying.

The problem is still unsolved. At the end of the nineteenth century explanations were suggested that fit it with the period's mythology: the children must have come from Mars, which was believed to be habitable, and the weakness of the sun's rays on that planet must have given them their green coloring. But we now know that Mars, like the moon, is almost without atmosphere and that no life, human or other, is possible there. Besides, it would be hard to imagine how a storm or typhoon that rose on the planet Mars could carry two beings all the way to earth and put them in a cave!

The medical explanation for "blue babies" is well known, but it seems that there are also green children, whose coloring comes from another disease, rarer than the blue, of endocrine origin. It would be reassuring to think that someone, for reasons unknown, perhaps out of superstition, had abandoned the two green children in the cave. The problem is that no sign of such a disappearance could be traced in the hospitals at the time.

It is useless to insist on modern hypotheses that posit the existence of parallel worlds or the possibility of the fourth dimension's intervening. This is today's mythology, which perhaps does not correspond to reality any more than the belief in Mars' habitability that was so widespread in the nineteenth century.

The hypothesis of a subterranean world is not absurd *a priori,* but it completely lacks proof. There is nothing to make us believe that there are inhabited caves at great depths. This hypothesis is raised periodically, but, on the basis of what we know about the earth's crust, it seems unfounded.

It is possible that surprises may be in store in this area and that the large numbers of legends and traditions about underground worlds—among which the Scandinavian tradition of *Hadding Land,* or hidden land, is especially detailed—may correspond to some reality. But in the current state of things, this seems highly unlikely.

There are many other hypotheses, including one that relates to the various hypotheses of this book: that the presence of these children was probably an experiment designed to provoke reactions in human beings. If this was the case, then it did not provoke very many. When the facts are truly disconcerting, people do not express much curiosity, and the account of the green children's story appears only in obscure compilations made by collectors of the bizarre.

Still it is worth our while to examine this story, just as it is worthwhile to examine all sorts of odd appearances in the framework of a series of experiments designed to measure the curiosity and intelligence of our civilization. Among such appearances, the Neanderthal man found recently by Heuvelmans and Sanderson deserves special mention. (Neither Heuvelmans nor Sanderson would accept my interpretation of their story.)

In early 1969, the two greatest specialists in "abominable snowmen" and other humanoids thought to be on the planet were traveling in the United States. While they were at a fair, they saw a booth that announced: *The oldest man of all time, enclosed in a block of ice.* They went in, more or less out of idle curiosity, and inside the block of ice they saw a Neanderthal man, bearing the mark of a bullet wound in his head. Needless to say, during the Neanderthal's time they hardly had firearms! The owner of the booth was fairly cooperative; he let them take photos and explained that he had bought the man in the ice in China.

Sanderson and Heuvelmans offered him huge sums of money for his attraction, which he at first refused and finally accepted, but when the two anthropologists returned with the money, supplied by the Smithsonian Institution, the man and his block of ice were gone.

Now it is not very easy to disappear completely, especially when you are being sought throughout the area by the FBI and when you are traveling with a large block of ice containing the remains of a Neanderthal man. It is not the sort of object that passes unnoticed.

Examination of the enlarged photos of the Neanderthal man show that the creature did indeed belong to an unknown species ancestral to man, very closely related to what we know of Neanderthal man. The bullet wound was unmistakable. The creature must have been brought into the United States by the same clandestine route by which drugs enter, and the Mafia was perhaps involved. For his part Heuvelmans received numerous threatening letters.

The block of ice and its contents were declared U.S. property, allowing the FBI officially to enter the search. Intermediaries between the crime syndicate and the police let it be known that if the FBI did not push its search too far in certain directions, the block of ice and its contents would be restored to the Smithsonian Institution for study, and that's where the matter rests for the moment.

In my opinion, however, we won't find the Neanderthal, any more than rats in experimental labyrinths find the piece of cheese after the experimenter lifts it into the third dimension with a hook. Heuvelmans and Sanderson told me that their creature must have been found floating in the Bering Strait, dead of a gunshot wound. According to them, on one side or another of the strait a tribe of Neanderthal men still is living. With all the respect I have for both anthropologists, I find it impossible to accept this hypothesis: both shores of the Bering Strait, Alaskan and Siberian, are armed territories that the Russians and Americans never stop surveying. There is radar on every square inch and security forces almost bump into each other there. One would be as likely to find a tribe of Neanderthals in the corridors of the Pentagon or in the cellars of the Kremlin.

In any case, until another event comes along to invalidate this hypothesis, I am sticking to the idea of a place where luminous demons, as well as pseudo-humans of the Cavendish type, green children, and Heuvelmans' Neanderthal are kept. They are withdrawn and put into circulation when it has been decided to make, and no doubt record, an experiment of our psychology and behavior, and then they are returned to the place from which they came. It seems to me that the majority of mysterious facts, especially these unexplained apparitions, constitute experiments of this type.

I would put into the same category those sudden shadows that fall in broad daylight when there is neither any cloudiness nor an eclipse. The classic case is the one that occurred on April 26, 1884, in Preston, England: toward noon, the sky became completely black, to the point that animals lay down and went to sleep. Twenty minutes later, the sun reappeared. We know of several hundred cases of this type, without having any explanation for them. It has been suggested that they are caused by thick clouds of smoke from forest fires, but generally there has been no sign of forest fire at the time of these incidents, and when there has been, these smoke clouds have never been observed between the spot where the fire took place and the place where the phenomenon occurred.

The strangest of these darkening phenomena occurred in London on August 19, 1763. The most amazing thing about this occurrence was that the shadows seemed to have been completely impenetrable by lantern or candlelight. If this was a case of smoke so thick that light could not pierce it, it would have left traces on objects, but it did not. This is disconcerting enough to make it worth adding to the number of these experiences.

It would also be tempting to include in this same collection appearances by abominable snowmen and abominable woodsmen, if they truly exist. Apparently serious observers declare that they have encountered hairy humanoids in the Soviet Union, the United States, and Tibet. Some specialists feel that these all belong to the same breed; Heuvelmans and Sanderson claim that it is a case of several different breeds, one of which is living in the United States. The "Big Feet," as they are called in the United States, are supposed to have shown themselves in highly populated areas. They are supposed to reproduce and to form, if not tribes, families or small groups. Perhaps, but it is hard to believe that, in a country as densely populated as the United States, there would not be more evidence.

The United States has something still better to offer us with the Flatwoods monster. In September, 1952, in the little Virginia village of Flatwoods, some children swore that they saw a monster come out of a vibrating red ball. A troop of children, led by a seventeen-year-old youth, who formed part of a voluntarily recruited police force, plunged into the forest. In the light of a flashlight beam the young policemen saw a creature twelve feet high, whose humanoid body was clothed in a rubber outfit that reflected the light and who was wearing a helmet. It had a red face, with two enormous orange-green eyes, and gave off an unpleasant odor. It moved, but without moving its feet; it seemed to glide instead. General panic ensued, affecting even the leader's dog, which was the first to take flight. They telephoned the sheriff, who did not find the monster but noted a disagreeable smell and unexplainable tracks.

All the children present, from twelve to seventeen years old, were questioned separately and in minute detail. Their stories agreed remarkably. Since the monster was not found, we cannot eliminate the possibility of a hoax. But fifteen years after these events occurred, none of the children, now grown up, had sold the story of this hoax to a magazine, a likely event had it been a hoax. Perhaps they really did see something.

The appearance of monsters of this type is noted periodically in every corner of the United States, which seems to have almost a monopoly on them. None of them resembles any other, which definitely rules out the possibility of an interplanetary civilization that might be paying us a visit. On the other hand, if these are experiments, it is logical enough that they would be different in each case.

With regard to abominable woodsmen, one was recently described as having escaped from his prison in Sumatra. He was remarkable for the hair that extended all over his body. We have been hearing about frequent examples of this sort of creature in British Columbia since 1884. They are seen, photographed, sold to circuses, all but made to walk in front of the television cameras. The zoologists assert that it is completely impossible for a creature the size of a gorilla to survive in our overpopulated human world, and we would certainly receive skeptically any news of such creatures living outside Paris. Nevertheless, apparently serious Americans have encountered humanoids as gigantic as they are hairy. For example, on July 23, 1963, in Oregon, at one o'clock in the morning, three people were driving in a car when a gray-haired humanoid twelve feet tall crossed the road. These witnesses are not alone. Again in Oregon, a couple from Portland were fishing in the area of the Lewis River when they saw a twelve-foot humanoid, standing on the bank. This one was wearing a Ku-Klux-Klan style hood, unless what they took for a hood was really its abundant mane. In August 1963, a newspaperman from the *Oregon Journal*, sent out to investigate, brought back excellent photos of huge footprints. According to these photographs, the weight of the creature who made them was some four hundred pounds. They looked more like giant human footprints than those of any known animal. Other footprints were photographed in the same month in the Lewis River area. These were even larger, and the creature who made them took thirty-foot steps and must have weighed more than seven hundred pounds.

As open as one's mind may be to the most extreme limits of credulity, it is difficult to believe that, in a country such as the United States, a reserve of humanoids this size could remain undiscovered. The area's forests are constantly surveyed by helicopters working on forest-fire prevention. If tribes of twelve-foot humanoids were walking about, they would have been noticed.

Until I receive more information, I will continue to believe that these animals are raised or manufactured elsewhere and deposited among us for experimental ends. I would tend to place them in the general category of *homunculi*, a term used in the Middle Ages to designate artificially manufactured humanoids.

I think that these creatures go far back in time. The ancient Greeks stressed the double aspect of the world in which they lived. They had creatures that looked like men but did not have an articulated language. They also had creatures, centaurs and satyrs, etc., who had bodies resembling those of both men and animals. Allusions to these two aspects of the real are so frequent in Greek literature that it is difficult to connect them only to mythology. For them, these creatures, half-men, half-beasts, were in no way divine but, on the contrary, completely physical. They did not vanish into smoke; they were not transparent; they could be seen, heard, and touched.

We therefore have grounds for assuming that the experimentation that led to the green children, to the Neanderthal man felled by gunshot, and to the twelve-foot apes who are roaming the United States in the twentieth century, perhaps began as far back as the origin of humanity and has been continued up to our time. None of these creatures seems capable, furthermore, of piloting interstellar machines or traveling in time. There are animals that have never been seen with a tool or with any manufactured object other than a helmet. They then must be brought to the place where they show themselves and taken back immediately, just as the experimenter retrieves the piece of cheese from the rat's labyrinth.

Up to now we have been speaking about giants. Small creatures also show themselves, but we must ask if this is the same phenomenon. No contemporary witness has described dwarfs, and we have no footprints. On the other hand, there are legends in every country referring to little men who live underground. They have even given their name to a metal: the word *cobalt* comes from *kobold,* one of the names that have been given them. But no one seems to have seen dwarfs since 1138, when one of them was captured in the cellar of a German monastery. This dwarf was completely black and spoke no language. They finally released him to see what he would do: he returned to the cellar where they had found him, lifted a stone, and then slipped into a tunnel where no one succeeded in following him. They sealed the tunnel with a cross, and that was where things were left.

The dwarf legend seems to have no connection

with the existence of Pygmies in Africa, about whom neither the Celts nor the American Indians knew. Nevertheless, a large number of traditions, as much among the Indians of both Americas as among the Celts and Europeans in general, speak of dwarfs who live beneath the earth. Margaret Murray even suggests that contact with the little people persisted into the contemporary epoch and that witchcraft is their ancient religion. In any case, we find green men, giants, and monsters of all sorts in our era, but no dwarfs. And if it is not correct to class dwarfs, elves, and other such creatures among the manifestations of extraterrestrial beings, neither may we class the various bizarre animals noted by Heuvelmans, Sanderson, and Prochnev among them either. The oceans and jungles are not fully explored, and for my part I am quite willing to admit that there can very conceivably be plesiosaurs in the Atlantic, sixty-foot-long snakes in South America, or dogs with two noses (the last two beasts having been observed by Fawcett before his mysterious disappearance). But we can fully admit the natural survival of these beasts without reference to the intervention of extraterrestrial experimenters.

On the other hand, I'd be rather inclined to think of "rapping spirits" as being another of these experiments of nonhuman origin. This phenomenon—technically, the *poltergeist* phenomenon—indisputably exists: blows are struck, objects are set in motion, and all in front of television cameras, reporters, and specialists in parapsychology. We have even seen a case in which a bottle of Javelle water, after being hung vertically in the air above the head of a distinguished parapsychologist, opened itself up, turned itself over, and drenched him—which makes one think that there are malicious Experimenters.

It has been established that the phenomenon occurs most often in the presence of young girls and boys of the age of puberty, without there being, it seems, any conscious responsibility on the part of the subject.

A great number of demented theories have attached themselves to poltergeists, the most amazing being that of a psychoanalyst who claims that they are ghosts, not of a personality, but of a complex. According to him, a complex can have such a strong existence of its own within the personality that it survives the death of the physical body. Poor Freud!

A hypothesis involving an Experimenter or a group of Experimenters who are producing these phenomena in order to record reactions seems considerably more plausible to me. We ourselves do this sort of thing in our psychological studies of animals.

Just as the existence of ghosts is indisputable, that of the poltergeists has been well established. We should note that they are always harmless and have absolutely never hurt anybody—with one recorded exception, in 1966. The BBC was trying to film some poltergeist pictures in an old house when a camera was pushed abruptly by invisible hands—the act was recorded and broadcast by another camera—and thrown down three flights of stairs. It barely missed one of the television reporters, who could easily have been killed. The case is unique, and if it was a question of a complex, it was a highly developed anti-television complex.

These details aside, the cases of the poltergeists resemble experiments made on laboratory animals that do not have enough intelligence or imagination to detect the experimenter. This is why I am inclined to class them in the same category as green people—with this difference, however. Although there is only one, well-established case of a green child, poltergeists show themselves more often, so much so that we can estimate the number of perfectly established cases at ten thousand.

An attempt has been made to classify all these apparitions by grouping together the humanoid apparitions on the one side and the non-humanoid on the other, but without much success, and I am inclined to place together the green children, the Neanderthals who are still walking about these days, the abominable snowmen and woodsmen, and the different non-humanoids.

The hypothesis of a series of experiments at least has the merit of not requiring an explanation for each step that squares with a more general hypothesis, like that of extraterrestrial or subterranean races. If it were a case of an extraterrestrial or subterranean race visiting our world, all the visitors would resemble each other, more or less. But there is no resemblance between the green children and the Flatwoods monster. We likewise note that none of these beings feels the need to utter any moral or religious message, and that they seldom allow themselves to be captured.

All these phenomena deserve serious study, using measuring instruments sensitive to all sorts of fields or radiations. Probably modern techniques are very largely sufficient, in the same way that Pasteur's technique was for spotting microbes, but it will take researchers of a rather special mentality. The idea that we are being observed by beings that we cannot see and that we are being manipulated by unknown forces is a typically paranoid idea. Pushed too far, the concept could lead the researcher to the asylum. But on the other hand, lacking some such belief, an investigator would be incapable of mounting the sort of experiments that would detect whether we are being observed or manipulated. This is a delicate balance, on the razor's edge.

It is obviously not possible simply to create such an observer, but we should note that many scientists have been eccentric and there will surely come a day when one eccentric will obstinately prove the existence of the Experimenters. Perhaps the day has already arrived, and we have a case of people being careful and prudent enough to avoid total ridicule by not publishing anything. General opinion will certainly change on this topic, just as it changed with regard to microbes. Semmelweiss was persecuted, Pasteur was fought, but modern microbiologists are receiving the Nobel Prize.

Without a doubt, the first researcher furnishing proof that we are being observed will be locked up. The second will have trouble, but his successors will probably create a new science that will appear as natural to future generations as microbiology. This science might study, in a rationalist light, the phenomena about which questions have been raised in this book, and perhaps other questions, such as demonic possession.

In the event—the certainty—that such studies go forward, we should be ready to answer questions as to the consequences of possible discoveries, for if we are being observed, we should show the observers that we are intelligent beings.

In all likelihood, the Intelligences may doubt that we have the quality—will think that our behavior, our migrations on weekends and in summertime, our wars and our concentration camps cannot be the activities of intelligent life.

An outside observer—even if the observers who are experimenting on us are much more intelligent than we, it is nonetheless true that they do remain *outside*—could very well think that our activities were solely due to multiple-conditioned reflexes such as those of the bees, the termites, and the ants. This is what Maurois indicated close to half a century ago in his *Fragments d'Une Histoire Universelle, 1992,* and his remarks are still thoroughly pertinent today. Even observers and manipulators much more intelligent than we run the risk of understanding nothing about our activities.

Perhaps, then, as we seek to find out about the Intelligences, to make contact, we should also look to our own actions.

10

And Today?

Looking at the contemporary situation, we come across some fascinating examples of the sort we have been discussing.

First, something that happened in Siberia on June 30, 1908. That night, over the Ienissei River, an explosion occurred that was more powerful than that of the atomic bombs dropped on Hiroshima and Nagasaki, rather more comparable in force to our heaviest hydrogen bombs. At the time, trails of light left by the trajectory of unidentified objects were observed in the sky, and an attempt was made to identify these trails with the path of an object that might have caused the explosion. This identification remains doubtful. But if these light trails really did have a connection with the object that exploded in 1908, recent calculations done with the aid of a computer prove that this object was executing maneuvers in altitude as well as in direction. After the explosion, seismic shock waves and electrostatic and electromagnetic disturbances were noted all over the globe, just as they would later be detected after thermonuclear explosions.

Beginning in 1927, several Soviet scientific expeditions explored the terrain of the explosion. They found none of the usual meteorite debris, but they did uncover a charred site with upturned trees and indisputable traces of radioactivity. In 1963 this terrain still had a radioactivity reading higher than the average of the region. The testimony of witnesses collected before World War I recounted a phenomenon singularly reminiscent of the mushroom cloud observed during atomic explosions. Some of these witnesses died a few years after the explosion from a disease with symptoms similar to those of the leukemia induced by atomic radiation.

What happened? More than eighty hypotheses have been put forward. I cite mine not because it is based on more proof than any of the others, but because it does *not* call on any extraterrestrial agents. At least, this should prove that I do not systematically try to introduce extraterrestrial agencies to explain isolated manifestations.

At that time, political exiles in Siberia were not shut up in concentration camps, but enjoyed considerable freedom. Some of them were certainly able to manufacture explosives. I imagine that a group of these deportees, in the course of research on radioactivity, discovered a much simpler method than our own to free nuclear energy. According to my analysis, this group probably tested the process by remote control in a kite-balloon. The explosion must have surpassed their expectations and destroyed them. Whole groups of convicts disappeared during this period without anyone's paying any attention.

Official Soviet scientists think that the explosion was caused by a collision between the earth and a comet, but this theory positively does not hold water, for as the comet approached the earth it would have been seen. Furthermore, at the time, people were very sensitive to this phenomenon as a result of the approach of Halley's Comet, which, according to certain astronomers, was going to destroy the world. An unknown comet, speeding toward the earth on a collision course, would have provoked a general panic. No such observation took place.

On the other hand, U.S. scientists claim that what was involved was a collision between the earth and a fairly large quantity of anti-matter, on the theory that the collision of matter and anti-matter involves total destruction of the matter–anti-matter mass, accompanied by a release of energy many time greater per pound of mass than in the most powerful hydrogen-bomb explosions. Moreover, we have succeeded in manufacturing small quantities of anti-matter in the laboratory, made up of negative nuclei and positrons that revolve around them. In particular we have obtained some atoms of antihelium 3.

The difficulty with this hypothesis is in assuming that the anti-matter could have attained a low enough altitude without disintegrating as a result of contact with earth's atmosphere, which, we are completely sure, is normally composed of matter.

Up to now, no one has been able to remove this objection, and this is where Alexander Kazantzev comes in. Kazantzev is, among other things, an author of science fiction (for which he is bitterly reproached, as if it were an obvious proof of mental debility), but he also is a weapons engineer. During World War II he headed an institute that developed new weapons that he himself went to try out on the front lines against the Nazis—which implies a certain mental balance. He is also a good chess player and has invented extremely ingenious problems, which implies a logical mind. I have met with him at great length two or three times; mentally and physically he is as solid as a rock. Kazantzev began as a metallurgist—he still uses his hands—and has a thorough knowledge of the most modern mathematical physics. He has several degrees, but both his feet are firmly on the ground. In short, he has impeccable credentials. Now, Kazantzev's hypothesis is that the 1908 explosion involved a spaceship come from some other place. He adds, as if to worsen his case, that before the catastrophe some members of the crew might have been able to escape and could still be among us.

Needless to say, he has been violently attacked. The fact remains that his theory is remarkably plausible. But where did this spaceship come from?

All modern research seems to prove irrefutably that there is no life anywhere in the solar system except on earth. It must therefore have been an interstellar spaceship, and it is possible to think that this spaceship was sent to bring to earth—or, better, to bring back to its star of origin—some experimental material concerning the research of which we are the subject and with which this book deals.

This particular aside, Kazantzev's hypothesis seems the most likely to me of all those that have been put forward, and I have carefully examined the eighty hypotheses recently published in the Soviet magazine *Piroda* and reproduced in *Planète*. Some are extraordinarily ingenious—for example, that the explosion involved an extremely powerful laser ray sent from another planet, the energy of which caused the earth's atmosphere to explode—but hardly plausible. Indeed, we can reject practically all of the hypotheses, except Kazantzev's, which, as fantastic as it may be, is superior to any of the others, though no debris of the mysterious object has been found.

What is generally forgotten in studies of this phenomenon is the one essential: knowing that there was a second act. On the night of February 9, 1913, strange objects entered our atmosphere. They did not explode, as in 1908. They did not fall as meteorites would have. *They left again.*

There is no doubt about their existence. Although the first observations were made by farmers and amateur astronomers, those that followed were made by professional astronomers, among them Professor C. A. Chant of the University of Toronto. For more than three minutes he observed luminous bodies traveling *in a group*—a first group of four objects, followed by a group of three, then by a group of two. Some of these objects were flying sufficiently low to cause "sonic booms" comparable to those produced by supersonic airplanes. Their flight appeared horizontal and their speed relatively slow, much slower than that of meteorites, which are measured in miles per second.

Another professional astronomer, W. F. Denning, wrote in the journal of the Royal Astronomical Society of Canada that it was like an express train, with lighted windows, in the sky. Sightings made from a ship permitted the precise observation that, having come from Canada, the objects flew over the Bermudas, then Brazil, then Africa, where, in the absence of qualified observers and observatories, they were lost.

The explanations that orthodox astronomers give for this event are absurd. They would like to have us believe that several groups of meteorites became natural satellites of the earth at the same time, though the most elementary calculation shows that this is a grossly improbable event. Besides, once a satellite has entered the atmosphere, it loses energy by friction and inevitably falls. None of these objects fell.

Moreover, we would have been able to calculate the orbit. If they were satellites, they would have had to reappear ninety-one minutes later. *No one ever saw them again.* Likewise, we should note that the objects themselves were not seen, just their lights. Side lights? Rocket jet flames? Plasma? Atmospheric fluorescence from the effect of a photonic propeller? No one is in a position to say. In any case, these objects made a low enough descent to produce "sonic booms" and then had enough energy to leave again. One hardly need recall that in 1913, no one on earth was in a position to launch rockets or artificial satellites.

The explanation that comes to my mind is that an extraterrestrial operation that failed disastrously in 1908 succeeded in 1913. Nothing similar has occurred since that date.

We can try to describe this operation in terms of a technology analogous to ours, but much more highly perfected. We could say that spaceships were set into orbit near the earth to transmit and receive specimens from our planet with the aid of a matter transmitter, and that they then departed. But we should not forget that interstellar spaceships and matter transmitters are part of our modern mythology, based on science fiction and comic strips, and that this mythology may appear as naive a few centuries hence as is, today, the idea of using wild birds to go to the moon. It is wiser simply to say that an operation originating outside the earth failed in 1908 and had much better success in 1913, and that this success undoubtedly has had direct or indirect consequences that we are not for the moment in a position to detect.

Mystics can assuredly connect the events of 1908 and 1913 with that of the year 1 in our era, which began with the appearance of a new, dazzling star in the sky before the birth of Christ. We would be better off simply saying, using the noted Soviet scientist I. S. Chklovski's word for it, that there were "miracles" in the year 1, in the year 1908, and in the year 1913. These miracles were accompanied by signs in the sky, but in none of these three cases do we know what happened.

Chklovski, in fact, was the first to suggest that we study "miracles" both on earth and in space, and there are several to be found even with a cursory search.

For instance, Japanese astronomers seem to have observed atomic explosions on Mars. Chklovski himself believes he has demonstrated that one of Mars' satellites, Phobos, is artificial. We have observed objects in the universe called—depending on their species—quasars, pulsars, and interlopers, which appear to emit modulated signals with fantastic energy. Arthur C. Clarke has noted that new stars can be seen in the constellation Auriga, appearing at short intervals and continuously coming nearer. He concludes from this that an interstellar war is in progress in this region and that the front is getting closer to us. We have also observed amazing jets of energy escaping from certain objects in the sky, such as the M.87 nebula. The incorrigible Clarke immediately points out that in his opinion this is either a work of art or a religious manifestation—like the flame at the Arch of Triumph! (And there is a certain beauty in thinking that there are beings who are burning ten suns per second in memory of a galactic unknown.)

It is likely that even before we can interpret signals from space, we will observe "miracles" proving that the Intelligences are operating on stars and perhaps on entire galaxies. For example, the most recent observations of gravitational waves show that matter in the sense in which it exists in our galaxy has been rearranged in a manner contrary to the natural laws we are familiar with. And if, as some think, the speed of gravitational waves is infinite, this rearrangement is always in process.

Of course, even if we succeed in scientifically proving the existence of the Intelligences in the cosmos, it will be much more difficult to prove that they intervene in our affairs, because they are very likely more subtle than the events of 1908 and 1913 would lead one to believe. It would obviously be very valuable to observe a contemporary intervention, and a detailed and methodical examination of the earth's history since 1913, the date of the last known spectacular manifestation, would perhaps uncover such an occurrence.

In any case, we would have to begin, it seems to me, by renouncing any moralizing attitude, any confusion between the Intelligences and a god of goodness and justice. It is useless, for example, to ask why the Intelligences have not prevented such and such a great catastrophe in our contemporary history; this is absolutely not their function. An entomologist observing a war between two anthills does not ask who began it or who is right: he observes. He sometimes extracts a few ants from one or the other camp and marks them so that he can follow the advance of the victorious army; but he is almost certain that the ants do not notice this intervention and that they know absolutely nothing about the marking techniques (and the marking is not, in any case, designed to insure victory for either one of the anthills).

We should begin simply by asking what great interventions there might have been since 1913. The Russian scientist Tsiolkovsky, the father of astronautics, thought that in the twentieth century interventions would be of a psychological order; in other words, that the Intelligences would intervene by telepathy. Interventions of this kind have been looked for, just as they have been sought in messages obtained by automatic writing from extraterrestrial signals.

I do not consider it out of the question that someone will be able to succeed. I no longer consider it out of the question that there have been people capable of unconsciously receiving telepathic transmissions—among such, Helen Blavatsky and Rudolph Steiner are perhaps the best known because of the religions they founded.

Indeed, automatic writing could easily be the key; there have been many amazing experiences with this skill. Perhaps the most extraordinary case is that of the American dentist John Newbrough Ballou, who was one of the first to experiment with laughing gas. Ballou found that, under the influence of this gas, he wrote automatically. He then bought a machine that had just been invented—a typewriter, the third or fourth to have been manufactured in the United States. With it he wrote a book entitled *Ohaspe,* far ahead of the learning of his time, and even clearly ahead of ours.

I know of numerous cases of this phenomenon

that I call "telepathy with the infinite." In a state of automatic writing, the subject begins to write a veritable encyclopedia of known and unknown knowledge, a compilation that often surpasses the level of humanity's learning at the time it is written. And I am convinced that a certain number of these encyclopedias are either the result of extraterrestrial interventions, or the unconscious reception of a course in galactic culture that was not intended for us.

At a less important level and in a more ordinary area, the same phenomenon occurs when people pick up radio broadcasts with the fillings in their teeth—a rare but thoroughly established phenomenon. In the same way, there must be people whose nervous systems form "printed circuits" and who pick up broadcasts of information that are not intended for us at all.

There is thus a kind of involuntary intervention by extraterrestrial beings. And when they realize this, perhaps they will send us devices so that we can receive these "courses in galactic culture" regularly, leading us to progress a good deal more quickly than when science was the only source of information. Fred Hoyle is absolutely convinced of this and believes in the imminent possibility of our being entered in a sort of cultural telephone book.

We can also see interventions in phenomena that are unexplainable for the time being but that have been observed with regard to satellites, particularly Telstar and Cosmos. After they stop transmitting, these satellites start to operate again, and one expert from NASA has declared: "It was just as if they had been disassembled and reassembled again." The hypothesis of Intelligences studying one of these satellites to ascertain, without attracting attention, the level of our technical knowledge is completely plausible. The day when we retrieve a satellite whose equipment has been improved, we will have proof of these interventions.

Generally speaking, I find it possible to hope that we are entering a new era in which interventions will be replaced by contacts: then we will be part of a galactic community.

With regard to power, we are coming closer to the Intelligences. We soon will be able to produce disturbances on our sun by bombarding it with a large enough number of hydrogen bombs. We can already send signals perceptible at dozens of light years, though the speed of light is too slow to make it worthwhile to build more powerful transmitters. However, if we discover particles or radiations that are conveyed at practically infinite speed, something that, contrary to what too many ignorant popularizers state, is not in the least incompatible with the theory of relativity, we will be able to send signals from one end to the other of the galaxy and perhaps to other galaxies.

Of course, if our entry into what the Soviet writer Efremov has called "the great ring of intelligence" requires considerable moral progress, then that entry is not exactly around the corner—not to mention the fact that it is likely that a civilization with a high level of power but no moral direction would destroy itself automatically.

Discussions on this subject, even among the greatest scientists, are not conspicuous for their independence of mind. For instance, scientists almost always define a civilization by the amount of energy it produces. Only Teilhard de Chardin, more poet than scientist, posed moral questions. Yet probably it is these very moral questions that have led to our being put in quarantine, and it is hard to explain the fact that there has been no widespread contact between ourselves and extraterrestrial beings, if not by such a quarantine.

When will this quarantine end? In the current state of our knowledge, no one can say. But when the day comes we will truly become civilized.

Index

Index